PRINCIPLES OF ZOOLOGY

Edited by
Katyayani Tatiparti

Delve Publishing

Principles of Zoology

Edited by Katyayani Tatiparti

Edition 2018

Delve Publishing

Published by

Delve Publishing LLC

708 3rd Avenue, 6th Floor

New York

NY 10801

United States of America

www.delvepublishing.com

ISBN: 978-1-68095-810-2 (Hardback)

Printed in United States of America

Library of Congress Control Number: 2016920693

About the Editor

Katyayani Tatiparti

Katyayani Tatiparti has a Master's degree in Pharmaceutical Sciences from JNTU, Hyderabad, India; with the topic of research for thesis being 'Self Micro Emulsifying Drug Delivery System for Carbamazcpinc (BCS Class II Drug)'. She also has a PG Diploma in Patents Law from NALSAR University, Hyderabad, India. She has been an IPR Professional and an Assistant Professor of Pharmaceutics. She is currently pursuing her second Master's degree in Pharmaceutical Sciences at Wayne State University, USA where she is pursuing research in cancer therapy.

Table of Contents

List of Abbreviations xi
Preface xiii

1 INTRODUCTION TO ZOOLOGY

Introduction 1
Pre-Scientific Zoology 2
Ancient Eastern Cultures 2
Ancient Greek Traditions 3
Aristotelian Zoology 3
Aristotle 3
Hellenistic zoology 4
Medieval and Islamic zoology 4
Renaissance and Early Modern 5
From Anatomy to Systematic Taxonomy 5
Impact of the Microscope 7
In advance of on the Origin of Species 8
History of Zoology Since 1859 9
Second half of Nineteenth Century 10
Twentieth Century 15
Twenty-first Century 20
References 21

2 SCIENTIFIC CLASSIFICATION IN ZOOLOGY

Introduction 23
Definition 24
Alpha Taxonomy and Beta Taxonomy 24
Microtaxonomy and Macrotaxonomy 25
Pre-Linnaean 26
Medieval 27
Modern System of Classification 28

Recent comprehensive classifications 30
Application 30
Author Citation (Zoology) 32
Rank Matters 32
Identity of the Author(S) 33
Examples to Illustrate Practical Use 34
Spelling of the Name of the Author 35
Inferred and Anonymous Authorships 36
Initials 36
Implications for Information Retrieval 36
Sensu Names 37
Phenetics 38
Difference from Cladistics 38
Phenetics Today 39
Databases 39
References 40

3 INVERTEBRATE ZOOLOGY

Introduction 46
Subdivisions 46
Entomology 47
In Pest Control 49
Taxonomic Specialization 50
Organizations 51
Museums 51
Africa 51
Europe 51
United States 52
Canada 53
Carcinology 53
Subfields 53
Crayfish 53
Names 54
Other Animals 54
Anatomy 54
Geographical Distribution and Classification 54
North America 56
Australia 56
New Zealand 56
Fossil Record 56
Crayfish Plague 57

Uses 57
Bait 58
Pets 58
Arachnology 59
Arachnology as a Science 60
Subdisciplines 60
Arachnological societies 60
Arachnological Journals 61
Popular Arachnology 61
Refernces 61

4 VERTEBRATE ZOOLOGY

Introduction 65
Ichthyology 66
Modern Publications 69
Organizations 70
Notable Ichthyologists 70
Paleoichthyologists 72
Fish 72
Evolution 73
Diversity 76
Anatomy 80
Hearing 84
Capacity for pain 84
Homeothermy 85
Reproductive System 86
Immune System 90
Conservation 91
Importance to Humans 92
Economic Importance 92
Terminology 95
Shoal or School 95
References 96

5 MAMMALOGY

Introduction 101
Mammal 103
Classification 104
Definitions 104
Class Mammalia 105

Molecular Classification of Placentals....106
Group I: Superorder Afrotheria....107
Statistics....108
Taxonomy and Phylogeny....108
Evolution from Amniotes....110
Anatomy and Morphology....114
Behavior....120
Locomotion....126
Mammals and Humans....130
References....133

6 BEHAVIORAL ECOLOGY

Competing for Resources....152
Sexual Selection....154
Mate Choice by Resources....154
Sexual Conflict....156
Parental Care and Family Conflicts....158
Mating Systems....163
Mating Systems with Male Parental Care....164
Social Behaviors....164
Kin Selection....165
Inclusive Fitness....165
Kin Recognition....165
Altruism and Conflict in Social Insects....168
Communication and Signalling....170
References....171

7 PRIMATOLOGY

Introduction....181
Sub-Disciplines....182
Western Primatology....182
Japanese Primatology....185
Origins....185
Primatology in Sociobiology....187
Women in Primatology....191
Academic Resources....192
References....192

8 CETALOGY

Introduction 195
Whale 195
Etymology and Definitions 197
Taxonomy and Evolution 198
Biology 201
Ecology 208
Interaction with Humans 210
Dolphin 217
Etymology 218
Taxonomy 219
Evolution 225
Biology 226
Behavior 229
Threats 235
Relationships with Humans 238
In Captivity 239
Consumption 243
Studying Cetaceans 245
Identifying Individuals 246
References 246

9 PALEOBIOLOGY

Introduction 259
Vertebrate Paleozoology 260
Quantitative Paleozoology 261
Conservation Biology 262
References 262

Index 263

List of Abbreviations

AFRAS	African Arachnological Society
AAS	American Arachnological Society
ASJ	Arachnological Society of Japan
ASA	Asian Society of Arachnology
ICZN Code	International Code of Zoological Nomenclature
ISA	International Society of Arachnology
IUCN	International Union for Conservation of Nature
BAS	British Arachnological Society
IAS	Iranian Arachnological Society
MNI	Minimum Number of Individuals
ESA	European Society of Arachnology
ESS	evolutionarily stable strategies
NISP	Number of Identified Species
ACE	The Associate Certified Entomologist

Preface

Zoology is the division of biology that deals with the animal kingdom. It's the scientific study of everything having to do with animals, just as botany is the scientific study of plants. Zoology is a huge field that covers the classification of every animal on earth as well as many broader fields of experimentation and inquiry related to animal life, and the field keeps expanding due to scientific advances that continue to open new areas of research. It is one of the branches of biology, and therefore it is also referred to as animal biology. There are several sub-branches within this field, including ethology, zoography, and anthrozoology. Additionally, zoologists often specialize in the study of specific types of animals. For instance, an ornithologist studies birds, while a mammologist studies mammals. As zoology is a very interdisciplinary subject, there are a number of related fields, including taxonomy, paleontology, and evolutionary biology. One of the most closely related fields to zoology is taxonomy, which is the practice of classifying different types of organisms according to shared characteristics. There are specific organizational structures that animals are put into, usually starting with superfamilies, and ending with subspecies. This field is also related to systematics, in which zoologists classify animals based on their genus or species. Another related field is evolutionary biology. When done in the context of zoology, this field entails studying how animals originally evolved and how they changed over time. This is often closely connected with paleozoology. Ecology and environmental studies are also related subjects, particularly zoography and ethology. This text will give you an overview of zoology.

Content Coverage

Chapters One and *Two* focus an overview of zoology and scientific classification in zoology, respectively. Over the 18th and 19th centuries, zoology became increasingly professional scientific disciplines.

Chapters Three and *Four* cover invertebrate zoology and vertebrate zoology, separately. Invertebrates include all animals exclusive of vertebrates, in an artificial division set up centuries ago, while vertebrate zoology is concerned with the study of animals with a backbone, such as fish, amphibians, reptiles, birds and mammals.

Chapter Five deals with the study of mammals – a class of vertebrates with characteristics such as homeothermic metabolism, fur, four-chambered hearts, and complex nervous systems.

Chapter Six focuses the study of the evolutionary basis for animal behavior due to ecological pressures. The value of a social behavior depends in part on the social behavior of an animal's neighbors.

Chapter Seven presents the study of primates that looks at the biological and psychological aspects of non-human primates. The focus is on studying the common links between humans and primates.

Chapters Eight and *Nine* offer information about cetalogy and paleobiology, separately. Cetalogy studies the approximately eighty species of whales, dolphins, and porpoise in the scientific order Cetacea, however paleobiology focuses on examining the origins and history of life on Earth through use of the fossil record.

CHAPTER 1

Introduction to Zoology

Chapter Contents

- **Introduction**
- **Pre-Scientific Zoology**
- **Aristotelian Zoology**
- **Renaissance And Early Modern**
- **History of Zoology Since 1859**

Zoology is the science of studying about the animals. The animal kingdom has fascinated mankind for centuries together. People try and understand the evolution of life, extinction and emergence of species to understand different biological cycles. Many scientists have published catalogues of their observation about the existence of animals and plants in deep seas and far-off lands. The exploration of nature still continues.

INTRODUCTION

The history of zoology before Charles Darwin's 1859 theory of evolution traces the organized study of the animal kingdom from ancient to modern times. Although the concept of *zoology* as a single coherent field arose much later, systematic study of zoology is seen in the works of Aristotle and Galen in the ancient Greco-Roman world. This work was developed in the Middle Ages by Islamic medicine and scholarship, and in turn their work was extended by European scholars such as Albertus Magnus.

During the European Renaissance and early modern period, zoological thought was revolutionized in Europe by a renewed interest in empiricism and the discovery of many novel organisms. Prominent in this movement were the anatomist Vesalius and the physiologist William Harvey, who used experimentation and careful observation, and naturalists such as Carl Linnaeus and Buffon who began to classify the diversity of life and the fossil record, as well as the development and behavior of organisms. Microscopy revealed the previously unknown world of microorganisms, laying the groundwork for cell theory. The growing importance of natural theology, partly a response to the rise of mechanical philosophy, encouraged the growth of natural history (although it entrenched the argument from design).

Over the 18th and 19th centuries, zoology became increasingly professional scientific disciplines. Explorer-naturalists such as Alexander von Humboldt investigated the interaction between organisms and their environment, and the ways this relationship depends on geography—laying the foundations for biogeography, ecology and ethology. Naturalists began to reject essentialism and consider the importance of extinction and the mutability of species. Cell theory provided a new perspective on the fundamental basis of life. These developments, as well as the results from embryology and paleontology, were synthesized in Charles Darwin's theory of evolution by natural selection. In 1859, Darwin placed the theory of organic evolution on a new footing, by his discovery of a process by which organic evolution can occur, and provided observational evidence that it had done so.

PRE-SCIENTIFIC ZOOLOGY

An ancient Egyptian plows his fields with a pair of oxen, used as beasts of burden and a source of food.

The earliest humans must have had and passed on knowledge about animals to increase their chances of survival. This may have included unsystematic knowledge of human and animal anatomy and aspects of animal behavior (such as migration patterns). People learnt more about animals with the Neolithic Revolution about 10,000 years ago. Humans domesticated animals as people became pastoralists and then farmers instead of hunter-gatherers in civilisations such as those of ancient Egypt.

Ancient Eastern Cultures

The ancient cultures of Mesopotamia, the Indian subcontinent, and China, among others, produced renowned surgeons and students of the natural sciences such as Susruta and Zhang

Zhongjing, reflecting independent sophisticated systems of natural philosophy. Taoist philosophers, such as Zhuangzi in the 4th century BC, expressed ideas related to evolution, such as denying the fixity of biological species and speculating that species had developed differing attributes in response to differing environments. The ancient Indian Ayurveda tradition independently developed the concept of three humours, resembling that of the four humours of ancient Greek medicine, though the Ayurvedic system included further complications, such as the body being composed of five elements and seven basic tissues. Ayurvedic writers also classified living things into four categories based on the method of birth (from the womb, eggs, heat & moisture, and seeds) and explained the conception of a fetus in detail. They also made considerable advances in the field of surgery, often without the use of human dissection or animal vivisection. One of the earliest Ayurvedic treatises was the *Sushruta Samhita*, attributed to Sushruta in the 6th century BC. It was also an early materia medica, describing 700 medicinal plants, 64 preparations from mineral sources, and 57 preparations based on animal sources. However, the roots of modern zoology are usually traced back to the secular tradition of ancient Greek philosophy.

Ancient Greek Traditions

The pre-Socratic philosophers asked many questions about life but produced little systematic knowledge of specifically zoological interest—though the attempts of the atomists to explain life in purely physical terms would recur periodically through the history of zoology. However, the medical theories of Hippocrates and his followers, especially humorism, had a lasting impact.

ARISTOTELIAN ZOOLOGY

Aristotle

Aristotle reported correctly that electric rays were able to stun their prey.

The philosopher Aristotle created the science of biology, basing its theory on both his metaphysical principles and on observation. He proposed theories for the processes of metabolism, temperature regulation, information processing, embryonic development and inheritance. He made detailed observations of nature, especially the habits and attributes of animals in the sea at Lesbos. He classified 540 animal species, and dissected at least 50.

Aristotle, and nearly all Western scholars after him until the 18th century, believed that creatures were arranged in a graded scale of perfection rising from plants on up to humans: the *scala naturae* or Great Chain of Being.

Hellenistic zoology

A few scholars in the Hellenistic period under the Ptolemies—particularly Herophilus of Chalcedon and Erasistratus of Chios—amended Aristotle's physiological work, even performing experimental dissections and vivisections. Claudius Galen became the most important authority on medicine and anatomy. Though a few ancient atomists such as Lucretius challenged the teleological Aristotelian viewpoint that all aspects of life are the result of design or purpose, teleology (and after the rise of Christianity, natural theology) remained central to biological thought until the 18th and 19th centuries.

Medieval and Islamic zoology

Frederick II's influential *De arte venandi* explored bird morphology.

The decline of the Roman Empire led to the disappearance or destruction of much knowledge, though physicians still incorporated many aspects of the Aristotelian tradition into training and practice. In Byzantium and the Islamic world, many of Aristotle's works were translated into Arabic and commented upon by scholars such as Avicenna and Averroes.

Medieval Muslim physicians, scientists and philosophers made significant contributions to zoological knowledge between the 8th and 13th centuries during the Islamic Golden Age. The Afro-Arab scholar al-Jahiz (781–869) described early evolutionary ideas such as the struggle for existence. He also introduced the idea of a food chain, and was an early adherent of environmental determinism.

During the High Middle Ages, a few European scholars such as Hildegard of Bingen, Albertus Magnus and Frederick II expanded the natural history canon. Magnus's *De animalibus libri XXVI* was one of the most extensive studies of zoological observation published before modern times.

RENAISSANCE AND EARLY MODERN

From Anatomy to Systematic Taxonomy

Conrad Gesner (1516–1565). His *Historiae animalium* is considered the beginning of modern zoology.

The Renaissance was the age of collectors and travellers, when many of the stories were actually demonstrated as true when the living or preserved specimens were brought to Europe. Verification by collecting of things, instead of the accumulation of anecdotes, then became more common, and scholars developed a new faculty of careful observation. The Renaissance brought expanded interest in both empirical natural history and physiology. In 1543, Andreas Vesalius inaugurated the modern era of Western medicine with his seminal human anatomy treatise *De humani corporis fabrica*, which was based on dissection of corpses. Vesalius was the first in a series of anatomists who gradually replaced scholasticism with empiricism in physiology and medicine, relying on first-hand experience rather than authority and abstract reasoning. Bestiaries—a genre that combines both the natural and figurative knowledge of animals—also became more sophisticated. Conrad Gessner great zoological work, *Historiae animalium*, appeared in four volumes, 1551-1558, at Zürich, a fifth being issued in 1587. His works were the starting-point of modern zoology. Other major works were produced by William Turner, Pierre Belon, Guillaume Rondelet, and Ulisse Aldrovandi. Artists such as Albrecht Dürer and Leonardo da Vinci, often working with naturalists, were also interested in the bodies of animals and humans, studying physiology in detail and contributing to the growth of anatomical knowledge. In the 17th century, the enthusiasts of the new sciences, the investigators of nature by means of observation and experiment, banded themselves into academies or societies for mutual support and discourse. The first founded of surviving European academies, the Academia Naturae Curiosorum (1651) especially confined itself to the description and illustration of the structure of plants and animals; eleven years later (1662) the Royal Society of London was incorporated by royal charter, having existed without a name or fixed organisation for seventeen years previously (from 1645). A little later the Academy of Sciences of Paris was established by Louis XIV, later still the Royal Society of Sciences in Uppsala was founded. Systematizing, naming and classifying dominated zoology throughout much of the 17th and 18th centuries. Carl Linnaeus published a basic taxonomy for the natural world in 1735 (variations of which have been in use ever since), and in the 1750s introduced scientific names for all his species. While Linnaeus conceived of species as unchanging parts of a designed hierarchy, the other great naturalist of the 18th century, Georges-Louis Leclerc, Comte de Buffon, treated species as artificial categories and living forms as malleable—even suggesting the possibility of common descent. Though he was writing in an era before evolution existed, Buffon is a key figure in the history of evolutionary thought; his "transformist" theory would influence the evolutionary theories of both Jean-Baptiste Lamarck and Charles Darwin. Before the Age of Exploration, naturalists had little idea of the sheer scale of biological diversity. The discovery and description of new species and the collection of specimens became a passion of scientific gentlemen and a lucrative enterprise for entrepreneurs; many naturalists traveled the globe in search of scientific knowledge and adventure.

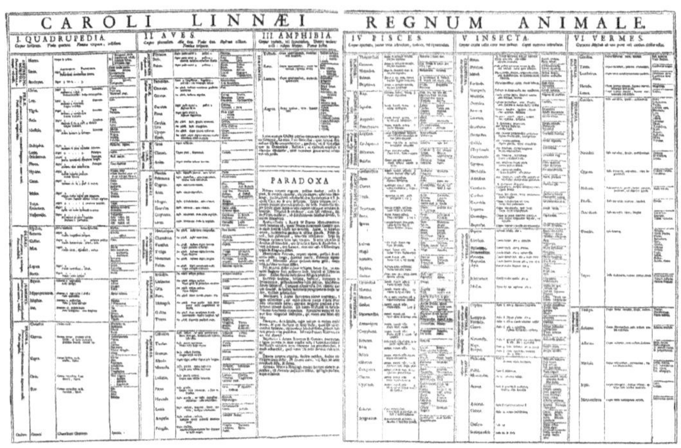

Table of the Animal Kingdom ("Regnum Animale") from the 1st edition of Linnaeus' *Systema Naturae* (1735).

Extending the work of Vesalius into experiments on still living bodies (of both humans and animals), William Harvey investigated the roles of blood, veins and arteries. Harvey's *De motu cordis* in 1628 was the beginning of the end for Galenic theory, and alongside Santorio Santorio's studies of metabolism, it served as an influential model of quantitative approaches to physiology.

Impact of the Microscope

In the early 17th century, the micro-world of zoology was just beginning to open up. A few lensmakers and natural philosophers had been creating crude microscopes since the late 16th century, and Robert Hooke published the seminal *Micrographia* based on observations with his own compound microscope in 1665. But it was not until Antony van Leeuwenhoek's dramatic improvements in lensmaking beginning in the 1670s—ultimately producing up to 200-fold magnification with a single lens—that scholars discovered spermatozoa, bacteria, infusoria and the sheer strangeness and diversity of microscopic life. Similar investigations by Jan Swammerdam led to new interest in entomology and built the basic techniques of microscopic dissection and staining.

Debate over the flood described in the Bible catalyzed the development of paleontology; in 1669 Nicholas Steno published an essay on how the remains of living organisms could be trapped in layers of sediment and mineralized to produce fossils. Although Steno's ideas about fossilization were well known and much debated among natural philosophers, an organic origin for all fossils would not be accepted by all naturalists until the end of the 18th century due to philosophical and theological debate about issues such as the age of the earth and extinction.

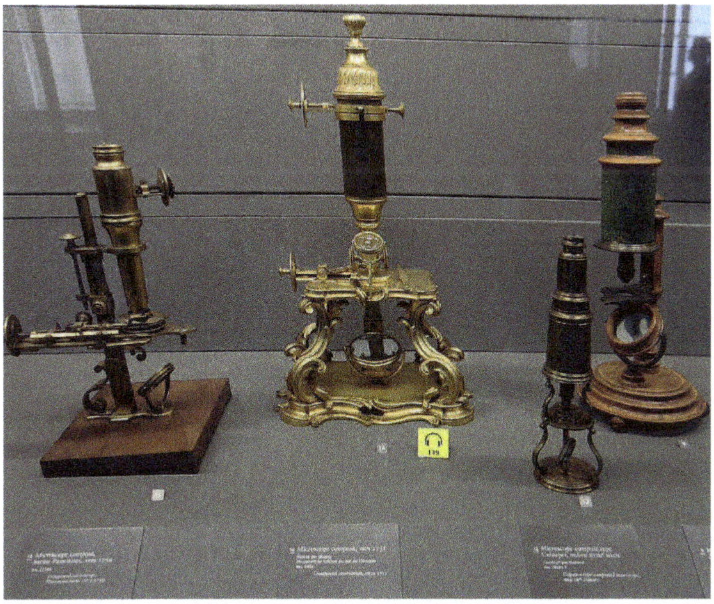

18th century microscopes from the Musée des Arts et Métiers, Paris.

Advances in microscopy also had a profound impact on biological thinking. In the early 19th century, a number of biologists pointed to the central importance of the cell. In 1838 and 1839, Schleiden and Schwann began promoting the ideas that (1) the basic unit of organisms is the cell and (2) that individual cells have all the characteristics of life, though they opposed the idea that (3) all cells come from the division of other cells. Thanks to the work of Robert Remak and Rudolf Virchow, however, by the 1860s most biologists accepted all three tenets of what came to be known as cell theory.

In advance of On the Origin of Species

Up through the 19th century, the scope of zoology was largely divided between physiology, which investigated questions of form and function, and natural history, which was concerned with the diversity of life and interactions among different forms of life and between life and non-life. By 1900, much of these domains overlapped, while natural history (and its counterpart natural philosophy) had largely given way to more specialized scientific disciplines—cytology, bacteriology, morphology, embryology, geography, and geology. Widespread travel by naturalists in the early-to-mid-19th century resulted in a wealth of new information about the diversity and distribution of living organisms. Of particular importance was the work of Alexander von Humboldt, which analyzed the relationship between organisms and their environment (i.e., the domain of natural history) using the quantitative approaches of natural philosophy (i.e., physics and chemistry). Humboldt's work laid the foundations of biogeography and inspired several generations of scientists.

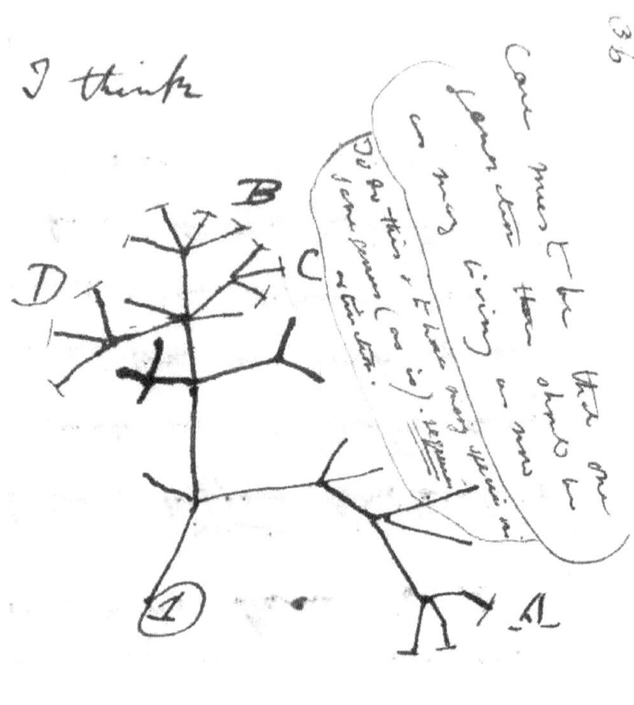

Charles Darwin's first sketch of an evolutionary tree from his *First Notebook on Transmutation of Species* (1837)

The emerging discipline of geology also brought natural history and natural philosophy closer together; Georges Cuvier and others made great strides in comparative anatomy and paleontology in the late 1790s and early 19th century. In a series of lectures and papers that made detailed comparisons between living mammals and fossil remains Cuvier was able to establish that the fossils were remains of species that had become extinct—rather than being remains of species still alive elsewhere in the world, as had been widely believed. Fossils discovered and described by Gideon Mantell, William Buckland, Mary Anning, and Richard Owen among others helped establish that there had been an 'age of reptiles' that had preceded even the prehistoric mammals. These discoveries captured the public imagination and focused attention on the history of life on earth.

The most significant evolutionary theory before Darwin's was that of Jean-Baptiste Lamarck; based on the inheritance of acquired characteristics (an inheritance mechanism that was widely accepted until the 20th century), it described a chain of development stretching from the lowliest microbe to humans. The British naturalist Charles Darwin, combining the biogeographical approach of Humboldt, the uniformitarian geology of Lyell, Thomas Malthus's writings on population growth, and his own morphological expertise, created a more successful evolutionary theory based on natural selection; similar evidence led Alfred Russel Wallace to independently reach the same conclusions. Charles Darwin's early interest in nature led him on a five-year voyage on HMS *Beagle* which established him as an eminent geologist whose observations and theories supported Charles Lyell's uniformitarian ideas, and publication of his journal of the voyage made him famous as a popular author. Puzzled by the geographical distribution of wildlife and fossils he collected on the voyage, Darwin investigated the transmutation of species and conceived his theory of natural selection in 1838. Although he discussed his ideas with several naturalists, he needed time for extensive research and his geological work had priority. He was writing up his theory in 1858 when Alfred Russel Wallace sent him an essay which described the same idea, prompting immediate joint publication of both of their theories. Darwin's *On the Origin of Species*, published on 24 November 1859, a seminal work of scientific literature, was to be the foundation of evolutionary biology.

HISTORY OF ZOOLOGY SINCE 1859

This article considers the history of zoology since the theory of evolution by natural selection proposed by Charles Darwin in 1859.

Charles Darwin gave new direction to morphology and physiology, by uniting them in a common biological theory: the theory of organic evolution. The result was a reconstruction of the classification of animals upon a genealogical basis, fresh investigation of the development of animals, and early attempts to determine their genetic relationships. The end of the 19th century saw the fall of spontaneous generation and the rise of the germ theory of disease, though the mechanism of inheritance remained a mystery. In the early 20th century, the rediscovery of Mendel's work led to the rapid development of genetics by Thomas Hunt Morgan and his students, and by the 1930s the combination of population genetics and natural selection in the "neo-Darwinian synthesis".

Second half of Nineteenth Century

Darwin and the Theory of Evolution

The 1859 publication of Darwin's theory in *On the Origin of Species by Means of Natural Selection, or the Preservation of Favoured Races in the Struggle for Life* is often considered the central event in the history of modern zoology. Darwin's established credibility as a naturalist, the sober tone of the work, and most of all the sheer strength and volume of evidence presented, allowed *Origin* to succeed where previous evolutionary works such as the anonymous *Vestiges of Creation* had failed. Most scientists were convinced of evolution and common descent by the end of the 19th century. However, natural selection would not be accepted as the primary mechanism of evolution until well into the 20th century, as most contemporary theories of heredity seemed incompatible with the inheritance of random variation.

Alfred Russel Wallace, following on earlier work by de Candolle, Humboldt and Darwin, made major contributions to zoogeography. Because of his interest in the transmutation hypothesis, he paid particular attention to the geographical distribution of closely allied species during his field work first in South America and then in the Malay archipelago. While in the archipelago he identified the Wallace line, which runs through the Spice Islands dividing the fauna of the archipelago between an Asian zone and a New Guinea/Australian zone. His key question, as to why the fauna of islands with such similar climates should be so different, could only be answered by considering their origin. In 1876 he wrote *The Geographical Distribution of Animals*, which was the standard reference work for over half a century, and a sequel, *Island Life*, in 1880 that focused on island biogeography. He extended the six-zone system developed by Philip Sclater for describing the geographical distribution of birds to animals of all kinds. His method of tabulating data on animal groups in geographic zones highlighted the discontinuities; and his appreciation of evolution allowed him to propose rational explanations, which had not been done before.

The scientific study of heredity grew rapidly in the wake of Darwin's *Origin of Species* with the work of Francis Galton and the biometricians. The origin of genetics is usually traced to the 1866 work of the monk Gregor Mendel, who would later be credited with the laws of inheritance. However, his work was not recognized as significant until 35 years afterward. In the meantime, a variety of theories of inheritance (based on pangenesis, orthogenesis, or other mechanisms) were debated and investigated vigorously.

In 1859, Charles Darwin placed the whole theory of organic evolution on a new footing, by his discovery of a process by which organic evolution can occur, and provided observational evidence that it had done so. This changed the attitudes of most exponents of the scientific method. Darwin's discoveries revolutionised the zoological and botanical sciences, by introducing the theory of evolution by natural selection as an explanation for the diversity of all animal and plant life. The subject-matter of this new science, or branch of biological science, had been neglected: it did not form part of the studies of the collector and systematist, nor was it a branch of anatomy, nor of the physiology pursued by medical men, nor again was it included in the field of microscopy and the cell theory. Almost a thousand years before Darwin, the Arab scholar Al-Jahiz (781 – 868) had already developed a rudimentary theory of natural selection , describing the Struggle for existence in his *Book of Animals* where he speculates on how environmental factors can affect the characteristics of species by forcing them to adapt and then passing on those new traits to future generations. However, his work had largely been forgotten, along with many other early advances of Arab scientists, and there is no evidence that his works were known to Darwin.

The area of biological knowledge which Darwin was the first to subject to scientific method and to render, as it were, contributory to the great stream formed by the union of the various branches, is that which relates to the breeding of animals and plants, their congenital variations, and the transmission and perpetuation of those variations. This branch of biological science may be called thremmatology - the science of breeding.Outside the scientific world, an immense mass of observation and experiment had grown up in relation to this subject. From the earliest times the shepherd, the farmer, the horticulturist, and the fancier had for practical purposes made themselves acquainted with a number of biological laws, and successfully applied them without exciting more than an occasional notice from the academic students of biology. Darwin made use of these observations and formulated their results to a large extent as the laws of variation and heredity. As the breeder selects a congenital variation which suits his requirements, and by breeding from the animals (or plants) exhibiting that variation obtains a new breed specially characterised by that variation, so in nature is there a selection amongst all the congenital variations of each generation of a species. This selection depends on the fact that more young are born than the natural provision of food will support. In consequence of this excess of births there is a struggle for existence and a survival of the fittest, and consequently an ever-present necessarily acting selection, which either maintains accurately the form of the species from generation to generation or leads to its modification in correspondence with changes in the surrounding circumstances which have relation to its fitness for success in the struggle for life, structures to the service of the organisms in which they occur.

It cannot be said that previously to Darwin there had been any very profound study of teleology, but it had been the delight of a certain type of mind, that of the lovers of nature or naturalists par excellence as they were sometimes termed, to watch the habits of living animals and plants and to point out the remarkable ways in which the structure of each variety of organic life was adapted to the special circumstances of life of the variety or species. The astonishing colours and grotesque forms of some animals and plants which the museum zoologists gravely described without comment were shown by these observers of living nature to have their significance in the economy of the organism possessing them; and a general doctrine was recognized, to the effect that no part or structure of an organism is without definite use and adaptation, being designed by the Creator for the benefit of the creature to which it belongs, or else for the benefit, amusement or instruction of his highest creatureman. Teleology in this form of the doctrine of design was never very deeply rooted amongst scientific anatomists and systematists. It was considered permissible to speculate somewhat vaguely on the subject of the utility of this or that startling variety of structure; but few attempts, though some of great importance, were made systematically to explain by observation and experiment the adaptation of organic structures to particular purposes in the case of the lower animals and plants. Teleology had, indeed, an important part in the development of physiology - the knowledge of the mechanism, the physical and chemical properties, of the parts of the body of man and the higher animals allied to him. But, as applied to lower and more obscure forms of life, teleology presented almost insurmountable difficulties; and consequently, in place of exact experiment and demonstration, the most reckless though ingenious assumptions were made as to the utility of the parts and organs of lower animals.

Darwin's theory had as one of its results the reformation and rehabilitation of teleology. According to that theory, every organ, every part, colour and peculiarity of an organism, must either be of benefit to that organism itself or have been so to its ancestors: no peculiarity of structure or general conformation, no habit or instinct in any organism, can be supposed to exist for the benefit or amusement of another organism, not even for the delectation of man himself.

A very subtle and important qualification of this generalization has to be recognized (and was recognized by Darwin) in the fact that owing to the interdependence of the parts of the bodies of living things and their profound chemical interactions and peculiar structural balance (what is called organic polarity) the variation of one single part (a spot of colour, a tooth, a claw, a leaflet) may, and demonstrably does in many cases entail variation of other parts what are called correlated variations. Hence many structures which are obvious to the eye, and serve as distinguishing marks of separate species, are really not themselves of value or use, but are the necessary concomitants of less obvious and even altogether obscure qualities, which are the real characters upon which selection is acting. Such correlated variations may attain to great size and complexity without being of use. But eventually they may in turn become, in changed conditions, of selective value. Thus in many cases the difficulty of supposing that selection has acted on minute and imperceptible initial variations, so small as to have no selective value, may be got rid of. A useless correlated variation may have attained great volume and quality before it is (as it were) seized upon and perfected by natural selection. All organisms are essentially and necessarily built up by such correlated variations.

Necessarily, according to the theory of natural selection, structures either are present because they are selected as useful or because they are still inherited from ancestors to whom they were useful, though no longer useful to the existing representatives of those ancestors. Structures previously inexplicable were now explained as survivals from a past age, no longer useful though once of value. Every variety of form and colour was urgently and absolutely called upon to produce its title to existence either as an active useful agent or as a survival. Darwin himself spent a large part of the later years of his life in thus extending the new teleology.

The old doctrine of types, which was used by the philosophically minded zoologists (and botanists) of the first half of the 19th century as a ready means of explaining the failures and difficulties of the doctrine of design, fell into its proper place under the new dispensation. The adherence to type, the favourite conception. of the transcendental morphologist, was seen to be nothing more than the expression of one of the laws of thremmatology, the persistence of hereditary transmission of ancestral characters, even when they have ceased to be significant or valuable in the struggle for existence, whilst the so-called evidences of design which was supposed to modify the limitations of types assigned to Himself by the Creator were seen to be adaptations due to the selection and intensification by selective breeding of fortuitous congenital variations, which happened to prove more useful than the many thousand other variations which did not survive in the struggle for existence.

Thus not only did Darwins theory give a new basis to the study of organic structure, but, whilst rendering the general theory of organic evolution equally acceptable and necessary, it explained the existence of low and simple forms of life as survivals of the earliest ancestry of the more highly complex forms, and revealed the classifications of the systematist as unconscious attempts to construct the genealogical tree or pedigree of plants and animals. Finally, it brought the simplest living matter or formless protoplasm before the mental vision as the starting point whence, by the operation of necessary mechanical causes, the highest forms have been evolved, and it rendered unavoidable the conclusion that this earliest living material was itself evolved by gradual processes, the result also of the known and recognized laws of physics and chemistry, from material which we should call not living. It abolished the conception of life as an entity above and beyond the common properties of matter, and led to the conviction that the marvellous and exceptional qualities of that which we call living matter are nothing more nor less than an exceptionally complicated development of those chemical and physical properties which we recognize in a gradually ascending scale of evolution in the carbon compounds, containing nitrogen as well as oxygen, sulphur and hydrogen as constituent atoms of their enormous molecules. Thus mysticism was finally banished from the domain of biology, and zoology became one of the physical sciencesthe science which seeks to arrange and discuss the phenomena of animal life and form, as the outcome of the operation of the laws of physics and chemistry.

A subdivision of zoology which was at one time in favour is simply into morphology and physiology, the study of form and structure on the one hand, and the study of the activitics and functions of the forms and structures of the other. But a logical division like this is not necessarily conducive to the ascertainment and remembrance of the historical progress and present significance of the science. No such distinction of mental activities as that involved in the division of the study of animal life into morphology and physiology has ever really existed: the investigator of animal forms has never entirely ignored the functions of the forms studied by him, and the experimental inquirer into the functions and properties of animal tissues and organs has always taken very careful account of the forms of those tissues and organs. A more instructive subdivision must be one which corresponds to the separate currents of thought and mental preoccupation which have been historically manifested in western Europe in the gradual evolution of what is to-day the great river of zoological doctrine to which they have all been rendered contributory.

Cell theory, Embryology and Germ Theory

Cell theory led zoologists to re-envision individual organisms as interdependent assemblages of individual cells. Scientists in the rising field of cytology, armed with increasingly powerful microscopes and new staining methods, soon found that even single cells were far more complex than the homogeneous fluid-filled chambers described by earlier microscopists. Much of the research on cell reproduction came together in August Weismann's theory of heredity: he identified the nucleus (in particular chromosomes) as the hereditary material, proposed the distinction between somatic cells and germ cells (arguing that chromosome number must be halved for germ cells, a precursor to the concept of meiosis), and adopted Hugo de Vries's theory of pangenes. Weismannism was extremely influential, especially in the new field of experimental embryology.

Innovative laboratory glassware and experimental methods developed by Louis Pasteur and other biologists contributed to the young field of bacteriology in the late 19th century.

By the 1880s, bacteriology was becoming a coherent discipline, especially through the work of Robert Koch, who introduced methods for growing pure cultures on agar gels containing specific nutrients in Petri dishes. The long-held idea that living organisms could easily originate from nonliving matter (spontaneous generation) was attacked in a series of experiments carried out by Louis Pasteur, while debates over vitalism vs. mechanism (a perennial issue since the time of Aristotle and the Greek atomists) continued apace.

Physiology

Over the course of the 19th century, the scope of physiology expanded greatly, from a primarily medically oriented field to a wide-ranging investigation of the physical and chemical processes of life—including plants, animals, and even microorganisms in addition to man. *Living things as machines* became a dominant metaphor in biological (and social) thinking. Physiologists such as Claude Bernard explored (through vivisection and other experimental methods) the chemical and physical functions of living bodies to an unprecedented degree, laying the groundwork for endocrinology (a field that developed quickly after the discovery of the first hormone, secretin, in 1902), biomechanics, and the study of nutrition and digestion. The importance and diversity of experimental physiology methods, within both medicine and zoology, grew dramatically over the second half of the 19th century. The control and manipulation of life processes became a central concern, and experiment was placed at the center of biological education.

Twentieth Century

At the beginning of the 20th century, zoological research was largely a professional endeavour. Most work was still done in the natural history mode, which emphasized morphological and phylogenetic analysis over experiment-based causal explanations. However, anti-vitalist experimental physiologists and embryologists, especially in Europe, were increasingly influential. The tremendous success of experimental approaches to development, heredity, and metabolism in the 1900s and 1910s demonstrated the power of experimentation in biology. In the following decades, experimental work replaced natural history as the dominant mode of research.

Early 20th Century Work (variation and heredity)

After publication of his work *The Origin of Species*, Darwin became interested in the animal and plant mechanisms that confer advantages to individual members of a species. Much important work was done by Fritz Muller (*Für Darwin*), by Hermann Müller (*Fertilization of Plants by Insects*), August Weismann, Edward B. Poulton and Abbott Thayer. There was considerable progress during this period in the field that would become known as genetics, the laws of variation and heredity (originally known as *thremmatology*). The progress of microscopy gave a clearer understanding of the origin of the egg-cell and sperm-cell and the process of fertilization.

Mendel and Zoology

Mendel's experiments on cultivated varieties of plants were published in 1865, but attracted little notice until thirty-five years later, sixteen years after his death (see Mendelism). Mendel tried to gain a better understanding of heredity. His main experiments were with varieties of the edible pea. He chose a variety with one marked structural feature and crossed it with another variety in which that feature was absent. For example, he hybridized a tall variety, with a dwarf variety, a yellow-seeded variety with a green-seeded variety, and a smooth-seeded variety with a wrinkle-seeded variety. In each experiment, he concentrated on one character; after obtaining a first hybrid generation, he allowed the hybrids to self-fertilize, and recorded the number of individuals in the first, second, third, and fourth generations in which the chosen character appeared.

In the first hybrid generation, nearly all the individuals had the positive character, but in subsequent generations the positive character was not present in all individuals: half had the character and half did not. Thus the random pairing of two groups of reproductive cells yielded the proportion 1 PP, 2 PN, 1 NN, where P stands for the character and N for its absence - the character was present in three-quarters of the offspring and absent from a quarter. The failure of the character to distribute itself among all of the reproductive cells of a hybrid individual, and the limitation of its distribution to half only of those cells, prevents the swamping of a new character by interbreeding. The tendency of the proportions in the offspring is to give, in a series of generations, a reversion from the hybrid form PN to a race with the positive character and a race without it. This tendency favours the persistence of a new character of large volume suddenly appearing in a stock. The observations of Mendel thus favoured the view that the variations upon which natural selection acts are not small but large and discontinuous. However, it did not appear that large variations would be favoured any more than small ones, or that the eliminating action of natural selection upon an unfavourable variation could be checked.

Much confusion arose in discussions of this topic, because of defective nomenclature. Some authors used the word *mutation* only for large variations that appeared suddenly and that could be inherited, and *fluctuation* for small variations, whether they could be transmitted or not. Other authors used *fluctuation* only for small, acquired variations due to changes in food, moisture and other features of the environment. This kind of variation is not heritable, but the small variations Darwin thought important are. The best classification of the variations in organisms separates those that arise from congenital variations from those that arise from variations of the environment or the food-supply. The former are innate variations, the latter are “acquired variations”. Both innate and acquired variations include some that are more and some that are less obvious. There are slight innate variations in every new generation of every species; their greatness or smallness so far as human perception goes is not of much significance, their importance for the origin of new species depends on whether they are valuable to the organism in the struggle for existence and reproduction. An imperceptible physiological difference might be of selective value, and it might carry with it correlated variations that may or may not appeal to the human eye, but are of no selective value themselves.

The views of Hugo de Vries and others about the importance of saltatory variation, the soundness of which was still not generally accepted in 1910, may be gathered from the article Mendelism. A due appreciation of the far-reaching results of correlated variation must, it appeared, give a new and distinct explanation of large mutations, discontinuous variation, and saltatory evolution. The analysis of the specific variations of organic form to determine the nature and limitation of a single character, and whether two variations of a structural unit can blend when one is transmitted by the male parent and the other by the female, were yet to be determined. It was not clear whether absolute blending was possible, or whether all apparent blending was only a more-or-less minutely subdivided mosaic of non-combinable characters of the parents.

Another important development of Darwin’s conclusions deserves notice. The fact of variation was familiar: no two animals, even of the same brood, are alike. Jean-Baptiste Lamarck hypothesised that structural alterations acquired by a parent might be transmitted to the offspring, and as these are acquired by an animal or plant as a consequence of the action of the environment, the offspring would sometimes start with a greater fitness for those conditions than its parents started with. In turn, it would acquire a greater development of the same modification, which it would transmit to its offspring. Lamarck argued that, over several generations, a structural alteration might thus be acquired. The familiar illustration of Lamarck's hypothesis is that of the giraffe, whose long neck might, he suggested, was acquired by the efforts of a short-necked race of herbivores who stretched their necks to reach the foliage of trees in a land where grass was deficient, the effort producing a longer neck of each generation, which was then transmitted to the next. This process is known as 'direct adaptation'.

Such structural adaptations are acquired by an animal in the course of its life, but are limited in degree and rare, rather than frequent and obvious. Whether acquired characters could be transmitted to the next generation was a very different issue. Darwin excluded any assumption of the transmission of acquired characters. He pointed to the fact of congenital variation, and showed that congenital variations are arbitrary and non-significant.

Congenital Variation

At the beginning of the 20th century, the causes of congenital variation were obscure, although it was recognised that they were largely due to a mixing of the matter that constituted the fertilized germ or embryo-cell from two individuals. Darwin had shown that congenital variation was all-important. A popular illustration of the difference was this: a man born with four fingers only on his right hand might transmit this peculiarity to at least some of his children; but a man with one finger chopped off will produce children with five fingers. Darwin, influenced by some facts that seemed to favour the Lamarckian hypothesis, thought that acquired characters are sometimes transmitted, but did not consider that this mechanism was likely to be of great importance.

After Darwin's writings, there was an effort to find evidence for the transmission of acquired characters; ultimately, the Lamarckian hypothesis of transmission of acquired characters was not supported by evidence, and was dismissed. August Weismann argued from the structure of the egg-cell and sperm-cell, and from how and when they are derived in the growth of the embryo from the egg, that it was impossible that a change in parental structure could produce a representative change in the germ or sperm-cells.

The only evidence that seemed to support the Lamarckian hypothesis were the experiments of Charles Brown-Séquard, who produced epilepsy in guinea-pigs by bisection of the large nerves or spinal cord, which led him to believe that, in rare instances, the artificially-produced epilepsy and mutilation of the nerves was transmitted. The record of Brown-Séquard's original experiments was unsatisfactory, and attempts reproduce them were unsuccessful. Conversely, the vast number of experiments in the cropping of the tails and ears of domestic animals, as well as of similar operations on man, had negative results. Stories of tailess kittens, puppies, and calves, born from parents one of whom had been thus injured, are abundant, but failed to stand experimental examination.

Whilst evidence of the transmission of an acquired character proved wanting, the *a priori* arguments in its favour were recognized as flawed, and cases that appeared to favour the Lamarckian assumption were found to be better explained by the Darwinian principle. For example, the occurrence of blind animals in caves and in the deep sea was a fact that even Darwin regarded as best explained by the atrophy of the eye in successive generations through the absence of light and consequent disuse. However, it was suggested that this is better explained by natural selection acting on congenital fortuitous variations. Some animals are born with distorted or defective eyes. If a number of some species of fish are swept into a cavern, those with perfect eyes would follow the light and eventually escape, leaving behind those with imperfect eyes to breed in the dark place. In every succeeding generation this would be the case, and even those with weak but still seeing eyes would escape, until only a pure race of blind animals would be left in the cavern.

Transmission

It was argued that the elaborate structural adaptations of the nervous system that underlie instincts must have been slowly built up by the transmission to offspring of acquired experience. It seemed hard to understand how complicated instincts could be due to the selection of congenital variations, or be explained except by the transmission of habits acquired by the parent. However, imitation of the parent by the young account for some, and there are cases in which elaborate actions must be due to the natural selection of a fortuitously-developed habit. Such cases are the habits of 'shamming dead' and the combined posturing and colour peculiarities of certain caterpillars (Lepidoptera larvae) that cause them to resemble dead twigs or similar objects. The advantage to the caterpillar is that it escapes (say) a bird that would, were it not deceived, attack and eat it. Preceding generations of caterpillars cannot have acquired this habit of posturing by experience; either a caterpillar postures and escapes, or it does not posture and is eaten - it is not half eaten and allowed to profit by experience. Thus, we seem justified in assuming that there are many movements of stretching and posturing possible to caterpillars, that some had a fortuitous tendency to one position, some to another, and, that among all the variety of habitual movements, one is selected and perpetuated because it happened to make the caterpillar look more like a twig.

Record of the Past

Man, compared with other animals, has the fewest instincts and the largest brain in proportion to body size. He builds up, from birth onwards, his own mental mechanisms, and forms more of them, and takes longer in doing so, than any other animal. The later stages of evolution from ape-like ancestors have consisted in the acquisition of a larger brain and in the education of that brain. A new feature in organic development makes its appearance when we set out the facts of man's evolutionary history. This factor is the *record of the past*, which grows and develops by laws other than those affecting the perishable bodies of successive generations of mankind, so that man, by the interaction of the *record* and his educability, is subject to laws of development unlike those by which the rest of the living world is governed.

Ecology and Environmental Science

In the early 20th century, naturalists were faced with increasing pressure to add rigor and preferably experimentation to their methods, as the newly prominent laboratory-based biological disciplines had done. Ecology had emerged as a combination of biogeography with the biogeochemical cycle concept pioneered by chemists; field biologists developed quantitative methods such as the quadrat and adapted laboratory instruments and cameras for the field to further set their work apart from traditional natural history. Zoologists did what they could to mitigate the unpredictability of the living world, performing laboratory experiments and studying semi-controlled natural environments; new institutions like the Carnegie Station for Experimental Evolution and the Marine Biological Laboratory provided more controlled environments for studying organisms through their entire life cycles.

Charles Elton's studies of animal food chains was pioneering among the succession of quantitative methods that colonized the developing ecological specialties. Ecology became an independent discipline in the 1940s and 1950s after Eugene P. Odum synthesized many of the concepts of ecosystem ecology, placing relationships between groups of organisms (especially material and energy relationships) at the center of the field. In the 1960s, as evolutionary theorists explored the possibility of multiple units of selection, ecologists turned to evolutionary approaches. In population ecology, debate over group selection was brief but vigorous; by 1970, most zoologists agreed that natural selection was rarely effective above the level of individual organisms.

Classical genetics, the Modern Synthesis, and Evolutionary Theory

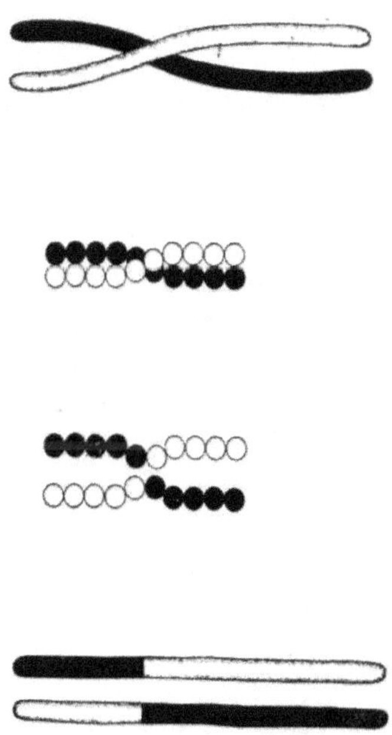

Thomas Hunt Morgan's illustration of crossing over, part of the Mendelian-chromosome theory of heredity

1900 marked the so-called *rediscovery of Mendel*: Hugo de Vries, Carl Correns, and Erich von Tschermak independently arrived at Mendel's laws (which were not actually present in Mendel's work). Soon after, cytologists (cell biologists) proposed that chromosomes were the hereditary material. Between 1910 and 1915, Thomas Hunt

Morgan and the "Drosophilists" in his fly lab forged these two ideas—both controversial—into the "Mendelian-chromosome theory" of heredity. They quantified the phenomenon of genetic linkage and postulated that genes reside on chromosomes like beads on string; they hypothesized crossing over to explain linkage and constructed genetic maps of the fruit fly *Drosophila melanogaster*, which became a widely used model organism.

Hugo de Vries tried to link the new genetics with evolution; building on his work with heredity and hybridization, he proposed a theory of mutationism, which was widely accepted in the early 20th century. Lamarckism also had many adherents. Darwinism was seen as incompatible with the continuously variable traits studied by biometricians, which seemed only partially heritable. In the 1920s and 1930s—following the acceptance of the Mendelian-chromosome theory— the emergence of the discipline of population genetics, with the work of R.A. Fisher, J.B.S. Haldane and Sewall Wright, unified the idea of evolution by natural selection with Mendelian genetics, producing the modern synthesis. The inheritance of acquired characters was rejected, while mutationism gave way as genetic theories matured.

In the second half of the century the ideas of population genetics began to be applied in the new discipline of the genetics of behavior, sociobiology, and, especially in humans, evolutionary psychology. In the 1960s W.D. Hamilton and others developed game theory approaches to explain altruism from an evolutionary perspective through kin selection. The possible origin of higher organisms through endosymbiosis, and contrasting approaches to molecular evolution in the gene-centered view (which held selection as the predominant cause of evolution) and the neutral theory (which made genetic drift a key factor) spawned perennial debates over the proper balance of adaptationism and contingency in evolutionary theory.

In the 1970s Stephen Jay Gould and Niles Eldredge proposed the theory of punctuated equilibrium which holds that stasis is the most prominent feature of the fossil record, and that most evolutionary changes occur rapidly over relatively short periods of time. In 1980 Luis Alvarez and Walter Alvarez proposed the hypothesis that an impact event was responsible for the Cretaceous–Paleogene extinction event. Also in the early 1980s, statistical analysis of the fossil record of marine organisms published by Jack Sepkoski and David M. Raup lead to a better appreciation of the importance of mass extinction events to the history of life on earth.

Twenty-first Century

Advances were made in analytical chemistry and physics instrumentation including improved sensors, optics, tracers, instrumentation, signal processing, networks, robots, satellites, and compute power for data collection, storage, analysis, modeling, visualization, and simulations. These technology advances allowed theoretical and experimental research including internet publication of zoological science. This enabled worldwide access to better measurements, theoretical models, complex simulations, theory predictive model experimentation, analysis, worldwide internet observational data reporting, open peer-review, collaboration, and internet publication.

REFERENCES

1. Magner, A History of the Life Sciences, pp 2–3
2. Needham, Joseph; Ronan, Colin Alistair (1995). The Shorter Science and Civilisation in China: An Abridgement of Joseph Needham's Original Text, Vol. 1. Cambridge University Press. p. 101. ISBN 0-521-29286-7.
3. Magner, A History of the Life Sciences, p. 6
4. Girish Dwivedi, Shridhar Dwivedi (2007). "History of Medicine: Sushruta – the Clinician – Teacher par Excellence" (PDF). National Informatics Centre. Retrieved 2008-10-08.
5. Magner, A History of the Life Sciences, pp 3–9
6. Magner, A History of the Life Sciences, pp 9–27
7. Mayr, The Growth of Biological Thought, pp 84–90, 135
8. Mason, A History of the Sciences, pp 41–44
9. Leroi, Armand Marie (2014). The Lagoon: How Aristotle Invented Science. Bloomsbury. pp. 370–373. ISBN 978-1-4088-3622-4.
10. Mayr, The Growth of Biological Thought, pp 201–202; see also: Lovejoy, The Great Chain of Being
11. Barnes, Hellenistic Philosophy and Science, p 383–384
12. Annas, Classical Greek Philosophy, p 252
13. Mayr, The Growth of Biological Thought, pp 91–94
14. Mehmet Bayrakdar, "Al-Jahiz And the Rise of Biological Evolutionism", The Islamic Quarterly, Third Quarter, 1983, London.
15. Paul S. Agutter & Denys N. Wheatley (2008). Thinking about Life: The History and Philosophy of Biology and Other Sciences. Springer. p. 43. ISBN 1-4020-8865-5.
16. Conway Zirkle (1941), Natural Selection before the "Origin of Species", Proceedings of the American Philosophical Society 84 (1): 71–123.
17. Frank N. Egerton, "A History of the Ecological Sciences, Part 6: Arabic Language Science - Origins and Zoological", Bulletin of the Ecological Society of America, April 2002: 142–146
18. Lawrence I. Conrad (1982), "Taun and Waba: Conceptions of Plague and Pestilence in Early Islam", Journal of the Economic and Social History of the Orient 25 (3), pp. 268–307 .
19. Albertus Magnus. On Animals: A Medieval Summa Zoologica. The Review of Metaphysics | December 01, 2001 | Tkacz, Michael W.
20. Mayr, The Growth of Biological Thought, pp 91–94:
21. Mayr, The Growth of Biological Thought, pp 166–171
22. Magner, A History of the Life Sciences, pp 80–83
23. Mayr, The Growth of Biological Thought, chapter 4

24. Mayr, The Growth of Biological Thought, chapter 7
25. See Raby, Bright Paradise
26. Magner, A History of the Life Sciences, pp 103–113
27. Magner, A History of the Life Sciences, pp 133–144
28. Rudwick, The Meaning of Fossils, pp 41–93
29. Sapp, Genesis, chapter 7; Coleman, Biology in the Nineteenth Century, chapters 2
30. Bowler, The Earth Encompassed, pp 204–211
31. Rudwick, The Meaning of Fossils, pp 112–113
32. Bowler, The Earth Encompassed, pp 211–220
33. Mayr, The Growth of Biological Thought, pp 343–357
34. Mayr, The Growth of Biological Thought, chapter 10: "Darwin's evidence for evolution and common descent"; and chapter 11: "The causation of evolution: natural selection"; Larson, Evolution, chapter 3
35. Desmond & Moore 1991, pp. 210, 284–285
36. Desmond & Moore 1991, pp. 263–274
37. van Wyhe 2007, pp. 184, 187
38. Beddall, B. G. (1968). "Wallace, Darwin, and the Theory of Natural Selection" (PDF). Journal of the History of Biology. 1 *(2): 261–323.* doi:10.1007/BF00351923.

CHAPTER 2

Scientific Classification in Zoology

Chapter Contents

- **Introduction**
- **Microtaxonomy and Macrotaxonomy**
- **Phenetics**

Taxonomy, in a broad sense, the science of classification, but more strictly the classification of living and extinct organisms—i.e., biological classification. The term is derived from the Greek taxis ("arrangement") and nomos ("law"). Taxonomy is, therefore, the methodology and principles of systematic botany and zoology and sets up arrangements of the kinds of plants and animals in hierarchies of superior and subordinate groups.

INTRODUCTION

Taxonomy (from Ancient Greek: τάξις taxis, "arrangement", and -νομία -nomia, "method") is the science of defining groups of biological organisms on the basis of shared characteristics and giving names to those groups. Organisms are grouped together into taxa (singular: taxon) and these groups are given a taxonomic rank; groups of a given rank can be aggregated to form a super group of lower rank, thus creating a taxonomic hierarchy. The Swedish botanist Carl Linnaeus is regarded as the father of taxonomy, as he developed a system known as Linnaean classification for categorization of organisms and binomial nomenclature for naming organisms.

With the advent of such fields of study as phylogenetics, cladistics, and systematics, the Linnaean system has progressed to a system of modern biological classification based on the evolutionary relationships between organisms, both living and extinct. An example of a modern classification is the one published in 2015 for all extant taxa (to the level of Order) by M. Ruggiero and co-workers.

Definition

The exact definition of taxonomy varies from source to source, but the core of the discipline remains: the conception, naming, and classification of groups of organisms. Two other terms are related to taxonomy, namely "systematics" and "classification"; their exact relationship to taxonomy also varies from source to source because the usage of the three terms in biology originated independently. As points of reference, recent definitions of taxonomy are presented below:

- Theory and practice of grouping individuals into species, arranging species into larger groups, and giving those groups names, thus producing a classification
- A field of science (and major component of systematics) that encompasses description, identification, nomenclature, and classification
- The science of classification, in biology the arrangement of organisms into a classification
- "The science of classification as applied to living organisms, including study of means of formation of species, etc."
- "The analysis of an organism's characteristics for the purpose of classification"
- "[Systematics] studies phylogeny to provide a pattern that can be translated into the classification and names of the more inclusive field of taxonomy" (listed as a desirable but unusual definition)

The varied definitions either place taxonomy as a sub-area of systematics (definition 2), invert that relationship (definition 6), or appear to consider the two terms synonymous. There is some disagreement as to whether biological nomenclature is considered a part of taxonomy (definitions 1 and 2), or a part of systematics outside taxonomy. For example, definition 6 is paired with the following definition of systematics that places nomenclature outside taxonomy:

- *Systematics*: "The study of the identification, taxonomy and nomenclature of organisms, including the classification of living things with regard to their natural relationships and the study of variation and the evolution of taxa".

Taxonomy, systematic biology, systematics, biosystematics, scientific classification, biological classification, phylogenetics: At various times in history, all these words have had overlapping meanings- sometimes the same, sometimes slightly different, but always overlapping and related.

The broadest meaning of "taxonomy" is used here. The word *taxonomy* was introduced in 1813 by Candolle, in his *Théorie élémentaire de la botanique.*

Alpha Taxonomy and Beta Taxonomy

The term "**alpha taxonomy**" is primarily used today to refer to the discipline of finding, describing, and naming taxa, particularly species. In earlier literature, the term had a different meaning, referring to morphological taxonomy, and the products of research through the end of the 19th century. William Bertram Turrill introduced the term "alpha taxonomy" in a series of papers published in 1935 and 1937 in which he discussed the philosophy and possible future directions of the discipline of taxonomy. ... there is an increasing desire amongst taxonomists to consider their problems from wider view-points, to investigate the possibilities of closer co-operation with their cytological, ecological and genetical colleagues and to acknowledge that some revision or expansion, perhaps of a drastic nature, of their aims and methods may be desirable ... Turrill (1935) has suggested that while accepting the older invaluable taxonomy, based on structure, and conveniently designated "alpha", it is possible to glimpse a far-distant taxonomy built up on as wide a basis of morphological and physiological facts as possible, and one in which "place is found for all observational and experimental data relating, even if indirectly, to the constitution, subdivision, origin and behaviour of species and other taxonomic groups". Ideals can, it may be said, never be completely realized. They have, however, a great value of acting as permanent stimulants, and if we have some, even vague, ideal of an "omega" taxonomy we may progress a little way down the Greek alphabet. Some of us please ourselves by thinking we are now groping in a "beta" taxonomy.

Turrill thus explicitly excludes from alpha taxonomy various areas of study that he includes within taxonomy as a whole, such as ecology, physiology, genetics, and cytology. He further excludes phylogenetic reconstruction from alpha taxonomy.

Later authors have used the term in a different sense, to mean the delimitation of species (not subspecies or taxa of other ranks), using whatever investigative techniques are available, and including sophisticated computational or laboratory techniques. Thus, Ernst Mayr in 1968 defined beta taxonomy as the classification of ranks higher than species.

An understanding of the biological meaning of variation and of the evolutionary origin of groups of related species is even more important for the second stage of taxonomic activity, the sorting of species into groups of relatives ("taxa") and their arrangement in a hierarchy of higher categories. This activity is what the term classification denotes; it is also referred to as **beta taxonomy**.

MICROTAXONOMY AND MACROTAXONOMY

How species should be defined in a particular group of organisms gives rise to practical and theoretical problems that are referred to as the species problem. The scientific work of deciding how to define species has been called microtaxonomy. By extension, macrotaxonomy is the study of groups at higher taxonomic ranks than species.

While some descriptions of taxonomic history attempt to date taxonomy to ancient civilizations, a truly scientific attempt to classify organisms did not occur until the 18th century. Earlier works were primarily descriptive, and focused on plants that were useful in agriculture or medicine. There are a number of stages in this scientific thinking. Early taxonomy was based on arbitrary criteria, the so-called "artificial systems", including Linnaeus's system of sexual classification. Later came systems based on a more complete consideration of the characteristics of taxa, referred to as "natural systems", such as those of de Jussieu (1789), de Candolle (1813) and Bentham and Hooker (1862–1863). These were pre-evolutionary in thinking. The publication of Charles Darwin's Origin of Species (1859) led to new ways of thinking about classification based on evolutionary relationships. This was the concept of phyletic systems, from 1883 onwards. This approach was typified by those of Eichler (1883) and Engler (1886–1892). The advent of molecular genetics and statistical methodology allowed the creation of the modern era of "phylogenetic systems" based on cladistics, rather than morphology alone.

Pre-Linnaean

Early Taxonomists

Taxonomy has been called "the world's oldest profession", and naming and classifying our surroundings has likely been taking place as long as mankind has been able to communicate. It would always have been important to know the names of poisonous and edible plants and animals in order to communicate this information to other members of the family or group.

Medicinal plant illustrations show up in Egyptian wall paintings from c. 1500 BC. The paintings clearly show that these societies valued and communicated the uses of different species, and therefore had a basic taxonomy in place.

Srimadbhagavata Purana

In Canto 3, chapter 10 of Srimadbhagavata purana 6 types of trees are recognised and they are:

- *Vanaspatis*-large trees that grow fruits without flowering.
- *Drumas*-large trees that bloom and give fruits.
- *Osadhis*-trees that die soon after they give fruits.
- *Latas*-creepers and tiny plants.
- *Viruts*-plants that grow as bushes.
- *Tvaksaras*-plants hollow inside with strong barks like bamboos.

Aristotle to Pliny the Elder

Organisms were first classified by Aristotle (Greece, 384–322 BC). He classified beings by their parts, or in modern terms attributes, such as having live birth, having four legs, laying eggs, having blood, or being warm-bodied. He divided all living things into two groups: plants and animals. Some of his groups of animals, such as *Anhaima* (animals without blood, translated as invertebrates) and *Enhaima* (animals with blood, roughly the vertebrates), as well as groups like the selachians and cetaceans, are still commonly used today. His student Theophrastus (Greece, 370–285 BC) carried on this tradition, and wrote a classification of some 500 plants called *Historia Plantarum*. Again, several plant groups currently still recognized can be traced back to Theophrastus, such as *Cornus*, *Crocus*, and *Narcissus*. The next major taxonomist was Pliny the Elder (Rome, 23–79 AD). His 160-volume work *Naturalis Historia* described many plants.

Medieval

Other pre-Linnaean taxonomists

It was about 1500 years later that taxonomic works became ambitious enough to replace the ancient texts. This is often credited to the development of sophisticated optical lenses, which allowed the morphology of organisms to be studied in much greater detail. One of the earliest authors to take advantage of this leap in technology was Andrea Cesalpino (Italy, 1519–1603), who is often referred to as "the first taxonomist". His magnum opus *De Plantis* came out in 1583, and described over 1500 plant species. Two large plant families that he first recognized are still in use today: the Asteraceae and Brassicaceae. Then in the 17th century John Ray (England, 1627–1705) wrote many important taxonomic works. Arguably his greatest accomplishment was *Methodus Plantarum Nova* (1682), in which he published details of over 18,000 plant species. At the time, his classifications were perhaps the most complex yet produced by any taxonomist, as he based his taxa on many combined characters. The next major taxonomic works were produced by Joseph Pitton de Tournefort (France, 1656–1708). His work from 1700, *Institutiones Rei Herbariae*, included over 9000 species in 698 genera, and directly influenced Linnaeus, as it was the text he used as a young student.

The Linnaean Era

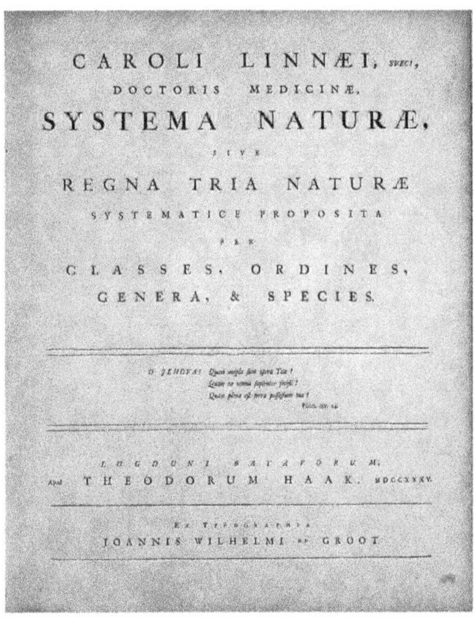
CAROLI LINNÆI, SVECI,
DOCTORIS MEDICINÆ,
SYSTEMA NATURÆ,
SIVE
REGNA TRIA NATURÆ
SYSTEMATICE PROPOSITA
PER
CLASSES, ORDINES,
GENERA, & SPECIES.

LUGDUNI BATAVORUM,
Apud THEODORUM HAAK. MDCCXXXV.

Ex Typographia
JOANNIS WILHELMI de GROOT

Title page of *Systema Naturae*, Leiden, 1735

The Swedish botanist Carl Linnaeus (1707–1778) ushered in a new era of taxonomy. With his major works *Systema Naturae* 1st Edition in 1735, *Species Plantarum* in 1753, and *Systema Naturae* 10th Edition,he revolutionized modern taxonomy. His works implemented a standardized binomial naming system for animal and plant species, which proved to be an elegant solution to a chaotic and disorganized taxonomic literature. Thus the Linnaean system was born, and is still used in essentially the same way today as it was in the 18th century. Currently, plant and animal taxonomists regard Linnaeus' work as the "starting point" for valid names (at 1753 and 1758 respectively). Names published before these dates are referred to as "pre-Linnaean", and not considered valid (with the exception of spiders published in *Svenska Spindlar*). Even taxonomic names published by Linnaeus himself before these dates are considered pre-Linnaean.

Modern System of Classification

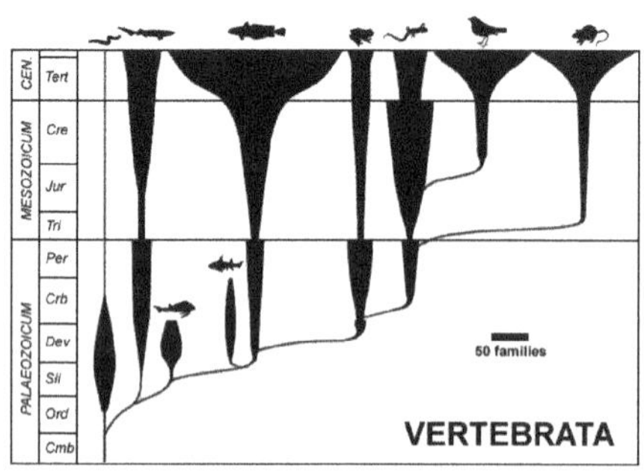

Evolution of the vertebrates at class level, width of spindles indicating number of families. Spindle diagrams are typical for Evolutionary taxonomy

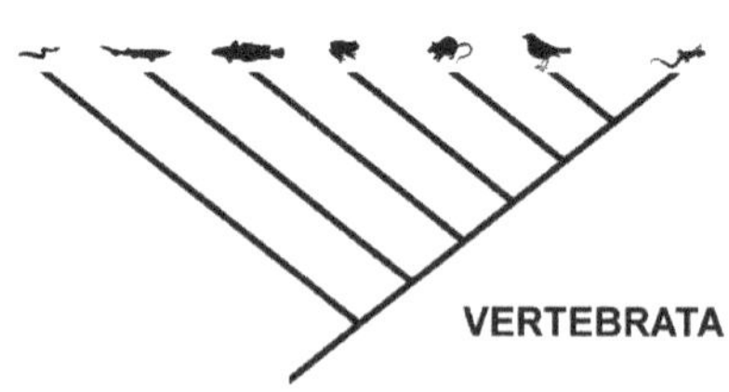

The same relationship, expressed as a cladogram typical for cladistics

Whereas Linnaeus classified for ease of identification, the idea of the Linnaean taxonomy as translating into a sort of dendrogram of the Animal- and Plant Kingdoms was formulated toward the end of the 18th century, well before *On the Origin of Species* was published. Among early works exploring the idea of a transmutation of species were Erasmus Darwin's 1796 *Zoönomia* and Jean-Baptiste Lamarck's *Philosophie Zoologique* of 1809. The idea was popularised in the Anglophone world by the speculative, but widely read *Vestiges of the Natural History of Creation*, published anonymously by Robert Chambers in 1844.

With Darwin's theory, a general acceptance that classification should reflect the Darwinian principle of common descent quickly appeared. Tree of Life representations became popular in scientific works, with known fossil groups incorporated. One of the first modern groups tied to fossil ancestors was birds. Using the then newly discovered fossils of *Archaeopteryx* and *Hesperornis*, Thomas Henry Huxley pronounced that they had evolved from dinosaurs, a group formally named by Richard Owen in 1842. The resulting description, that of dinosaurs "giving rise to" or being "the ancestors of" birds, is the essential hallmark of evolutionary taxonomic thinking. As more and more fossil groups were found and recognized in the late 19th and early 20th centuries, palaeontologists worked to understand the history of animals through the ages by linking together known groups. With the modern evolutionary synthesis of the early 1940s, an essentially modern understanding of the evolution of the major groups was in place. As evolutionary taxonomy is based on Linnaean taxonomic ranks, the two terms are largely interchangeable in modern use.

Since the 1960s, a trend called phylogenetic nomenclature (or cladism) has emerged, inspired by the cladistic method. The salient feature is arranging taxa in a hierarchical evolutionary tree, ignoring ranks. If a taxon includes all the descendants of some ancestral form, it is called monophyletic. Groups that have descendant groups removed from them (e.g. dinosaurs, with birds as offspring group) are termed paraphyletic, while groups representing more than one branch from the tree of life are called polyphyletic. The *International Code of Phylogenetic Nomenclature* or *PhyloCode* is intended to regulate the formal naming of clades. Linnaean ranks will be optional under the *PhyloCode*, which is intended to coexist with the current, rank-based codes.

Kingdoms and Domains

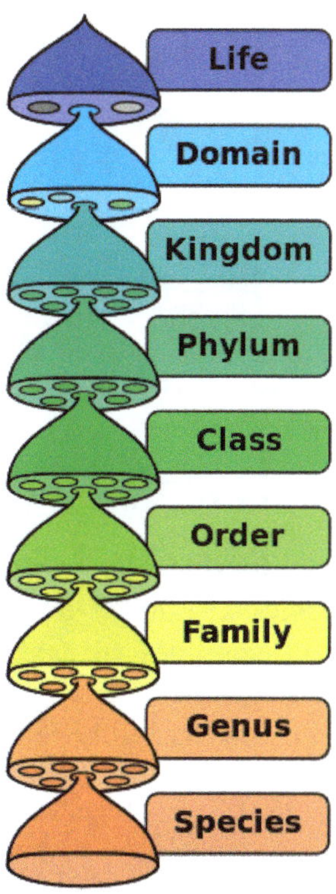

Modern Classification

Well before Linnaeus, plants and animals were considered separate Kingdoms. Linnaeus used this as the top rank, dividing the physical world into the plant, animal and mineral kingdoms. As advances in microscopy made classification of microorganisms possible, the number of kingdoms increased, five and six-kingdom systems being the most common.

Domains are a relatively new grouping. The three-domain system was first proposed in 1990, but not generally accepted until later. One main characteristic of the three-domain method is the separation of Archaea and Bacteria, previously grouped into the single kingdom Bacteria (a kingdom also sometimes called Monera). Consequently, the three domains of life are conceptualized as Archaea, Bacteria, and Eukaryota (comprising the nuclei-bearing eukaryotes). A small minority of scientists add Archaea as a sixth kingdom, but do not accept the domain method.

Thomas Cavalier-Smith, who has published extensively on the classification of protists, has recently proposed that the Neomura, the clade that groups together the Archaea and Eukarya, would have evolved from Bacteria, more precisely from Actinobacteria. His classification of 2004 treats the archaeobacteria as part of a subkingdom of the Kingdom Bacteria, i.e. he rejects the three-domain system entirely. Stefan Luketa in 2012 proposed a five "dominion" system, adding Prionobiota (acellular and without nucleic acid) and Virusobiota (acellular but with nucleic acid) to the traditional three domains.

Linnaeus 1735	Haeckel 1866	Chatton 1925	Copeland 1938	Whittaker 1969	Woese et al. 1990	Cavalier-Smith 1998
2 kingdoms	3 kingdoms	2 empires	4 kingdoms	5 kingdoms	3 domains	6 kingdoms
(not treated)	Protista	Prokaryota	Monera	Monera	Bacteria	Bacteria
					Archaea	
		Eukaryota	Protoctista	Protista	Eucarya	Protozoa
						Chromista
Vegetabilia	Plantae		Plantae	Plantae		Plantae
				Fungi		Fungi
Animalia	Animalia		Animalia	Animalia		Animalia

Recent comprehensive classifications

Partial classifications exist for many individual groups of organisms and are revised and replaced as new information becomes available, however comprehensive treatments of most or all life are rarer; two recent examples are that of Adl et al., 2012, which covers eukaryotes only with an emphasis on protists, and Ruggiero et al., 2015, covering both eukaryotes and prokaryotes to the rank of Order, although both exclude fossil representatives.

Application

Biological taxonomy is a sub-discipline of biology, and is generally practised by biologists known as "taxonomists", though enthusiastic naturalists are also frequently involved in the publication of new taxa. The work carried out by taxonomists is crucial for the understanding of biology in general. Two fields of applied biology in which taxonomic work is of fundamental importance are the studies of biodiversity and conservation. Without a working classification of the organisms in any given area, estimating the amount of diversity present is unrealistic, making informed conservation decisions impossible. As conservation becomes ever more politically important, it has been claimed that taxonomic work impacts not only the scientific community, but society as a whole.

Classifying Organisms

Biological classification is a critical component of the taxonomic process. As a result, it informs the user as to what the relatives of the taxon are hypothesized to be. Biological classification uses taxonomic ranks, including among others (in order from most inclusive to least inclusive): Domain, Kingdom, Phylum, Class, Order, Family, Genus, and Species.

Taxonomic Descriptions

Type specimen for *Nepenthes smilesii*, a tropical pitcher plant.

The "definition" of a taxon is encapsulated by its description and/or its diagnosis. There are no set rules governing the definition of taxa, but the naming and publication of new taxa is governed by sets of rules. In zoology, the nomenclature for the more commonly used ranks (superfamily to subspecies), is regulated by the *International Code of Zoological Nomenclature* (*ICZN Code*). In the fields of botany, phycology, and mycology, the naming of taxa is governed by the *International Code of Nomenclature for algae, fungi, and plants* (*ICN*).

The initial description of a taxon involves five main requirements:

- The taxon must be given a name based on the 26 letters in the Latin alphabet (a binomial for new species, or uninomial for other ranks).

- The name must be unique (i.e. not a homonym).
- The description must be based on at least one name-bearing type specimen.
- It should include statements about appropriate attributes either to describe (define) the taxon, and/or to differentiate it from other taxa (the diagnosis, *ICZN Code*, Article 13.1.1, *ICN*, Article 38). Both codes deliberately separate defining the content of a taxon (its circumscription) from defining its name.
- These first four requirements must be published in a work that is obtainable in numerous identical copies, as a permanent scientific record.

However, often much more information is included, like the geographic range of the taxon, ecological notes, chemistry, behavior, etc. How researchers arrive at their taxa varies: depending on the available data, and resources, methods vary from simple quantitative or qualitative comparisons of striking features, to elaborate computer analyses of large amounts of DNA sequence data.

AUTHOR CITATION (ZOOLOGY)

An "authority" may be placed after a scientific name. The authority is the name of the scientist who first validly published the name. For example, in 1758 Linnaeus gave the Asian elephant the scientific name *Elephas maximus*, so the name is sometimes written as "*Elephas maximus* Linnaeus, 1758". The names of authors are frequently abbreviated: the abbreviation *L.* is universally accepted for Linnaeus, and in botany there is a regulated list of standard abbreviations (see list of botanists by author abbreviation). The system for assigning authorities differs slightly between botany and zoology. However, it is standard that if a species' name or placement has been changed since the original description, the original authority's name is placed in parentheses.

In zoological nomenclature, author citation refers to listing the person (or team) who first makes a scientific name of a taxon available. This is done in a scientific publication while fulfilling the formal requirements under the International Code of Zoological Nomenclature, hereinafter termed "the Code". According to the Code, "the name of the author does not form part of the name of a taxon and its citation is optional, although customary and often advisable" (Article 51.1), however Recommendation 51A suggests: "The original author and date of a name should be cited at least once in each work dealing with the taxon denoted by that name. This is especially important in distinguishing between homonyms and in identifying species-group names which are not in their original combinations". For the purpose of information retrieval, the author citation and year appended to the scientific name, e.g. genus-species-author-year, genus-author-year, family-author-year, etc., is often considered a "de facto" unique identifier, although for a number of reasons discussed below, this usage may often be imperfect.

Rank Matters

The Code recognises three groups of names, according to rank:

- family-group names, at the ranks of superfamily, family, subfamily, tribe, subtribe (any rank below superfamily and above genus).

- genus-group names, at the ranks of genus and subgenus.
- species-group names, at the ranks of species and subspecies.

Within each group, the same authorship applies regardless of the taxon level to which the name (with, in the case of a family-group name, the appropriate ending) is applied. For example, the taxa that the red admiral butterfly can be assigned to:

- Family: Nymphalidae Swainson, 1827 so also
  - Subfamily: Nymphalinae Swainson, 1827 and
  - Tribe: Nymphalini Swainson, 1827
- Genus: *Vanessa* Fabricius, 1807 so also
  - Subgenus: *Vanessa (Vanessa)* Fabricius, 1807
- Species: *Vanessa atalanta* (Linnaeus, 1758) so also
  - Subspecies: *Vanessa atalanta atalanta* (Linnaeus, 1758)

(The parentheses around the author citation indicate that this was not the original taxonomic placement: in this case, Linnaeus published the name as

Papilio atalanta Linnaeus, 1758.)

Identity of the Author(S)

The identity of the author had long been a matter of dispute and of secondary importance. In the first attempt to provide international rules for zoological nomenclature in 1895, the author was defined as the author of the scientific description, and not as the person who provided the name (published or unpublished), as had been usual practice in various animal groups before. This had the result that in some disciplines, for example in malacology, most taxonomic names had to change their authorship because they had been attributed to other persons who never published a scientific work.

This new rule was however not sufficiently accurate and did not provide an exact guide, so that in the following decades taxonomic practice continued to diverge among disciplines and authors. The ambiguous situation led a member of the ICZN Commission in 1974 to provide an interpretation of Art. 50 of the second edition of the Code (effective since 1961), where the author had been defined as "the person who first publishes a scientific name in a way that satisfies the criteria of availability", an interpretation following which this should be seen as largely being restricted to providing a description or diagnosis.

Currently most (but not all) taxonomists accept this view and restrict authorship for a taxonomic name to the person who was responsible for having written the textual scientific content of the original description, or in other words, the visibly responsible person for having written down what the publisher finally published. The author of an image is not recognized as co-author of a name, even if the image was the only base provided for making the name available.

The author is usually the author of the work. But sometimes new zoological names were not established by that author.

If a true author of a written text is not directly recognisable in the original publication, she or he is not the author of a name (but the author of the work is). The text could actually be written by a different person. Some authors have copied text passages from unpublished sources without acknowledging them. In Art. 50.1.1 all these persons are excluded from the authorship of a name, if they were not explicitly mentioned in the work itself for being the responsible persons for making a name available.

Most taxonomists also accept Art. 50.1.1 that the author of a cited previously published source, from which text passages were copied, is not acknowledged as the author of a name.

Not all taxonomists seem to know this, and there are traditions in some animal groups (for example in fish nomenclature) where the "true" author of a work is still occasionally cited in the name of a species ("Walbaum (ex Klein), 1792").

In some cases the author of the description can differ from the author of the work. This must be explicitly indicated in the original publication, either by a general statement ("all zoological descriptions in this work were written by Smith"), or by an individual statement ("the following three descriptions were provided by Jiménez", "this name shall be attributed to me and Wang because she contributed to the description").

In the 1800s it was a usual style to eventually set an abbreviation of another author immediately below the text of the description or diagnosis to indicate authorship for the description. This is commonly accepted today: if the description is attributed to a different person, then that person is the author.

When the name of a different author was only set behind the new name in the headline (and not repeated below the diagnosis to indicate that that diagnosis had been written by that person), this was a convention to indicate authorship only for the new name and not for the description. These authorships for names are not covered by Art. 50.1 and are not accepted. Only authorship for the description is accepted.

Prior to 1900-1920 there were several different conventions concerning the authorships, every animal group had other traditions. This is why we frequently find other authors than today for zoological names in the early zoological literature. Art. 50.1 has been a quite successful model since it became commonly accepted in the mid-1900s. There is no need to research who the true author was, everyone including young and relatively unexperienced researchers can verify and determine the name of the author in the original work itself.

Examples to Illustrate Practical Use

In citing the name of an author, the surname is given in full, not abbreviated, with no mention of the first name(s). The date (true year) of publication in which the name was established is added, if desired with a comma between the author and date (the comma is not prescribed under the Code, it contains no additional information, however it is included in examples therein and also in the ICZN Official Lists and Indexes).

- *Balaena mysticetus* Linnaeus, 1758

the bowhead whale was described and named by Linnaeus in his *Systema Naturae* of 1758

- *Anser albifrons* (Scopoli, 1769)

the white-fronted goose was first described (by Giovanni Antonio Scopoli), as *Branta albifrons* Scopoli, 1769. It is currently placed in the genus *Anser*, so author and year are set in parentheses. The taxonomist who first placed the species in *Anser* is not recorded (and much less cited), the two different genus-species combinations are not regarded as synonyms.

An author can have established a name dedicated to oneself. This sounds unusual and is rare, and is against the unwritten conventions, but it is not restricted under the Code.

- *Xeropicta krynickii* (Krynicki, 1833)

a terrestrial gastropod from Ukraine was first described as *Helix krynickii* Krynicki, 1833, who originally attributed the name to another person Andrzejowski. But the description was written by Krynicki, and Andrzejowski had not published this name before.

Spelling of the Name of the Author

In a strict application of the Code the taxon name author string components "genus", "species" and "year" can only have one combination of characters. The major problem in zoology for consistent spellings of names is the author. The Code gives neither a guide nor a detailed recommendation.

Unlike in botany, it is not recommended to abbreviate the name of the author in zoology. If a name was established by more than three authors, only the first author may be given, followed by the term "et al." (and others).

There are no approved standards for spellings of authors in zoology, and unlike in botany no one has ever proposed such standards for zoological authors.

It is generally accepted that the name of the author shall be given in the nominative singular case if originally given in a different case, and that the name of the author should be spelled in Latin script. There are no commonly accepted conventions how to transcribe names of authors if given in non-Latin script.

It is also widely accepted that names of authors must be spelled with diacritic marks, ligatures, spaces and punctuation marks. The first letter is normally spelled in upper-case, however initial capitalization and usage of accessory terms can be inconsistent (e.g. de Wilde/De Wilde, d'Orbigny/D'Orbigny, Saedeleer/De Saedeleer, etc.). Co-authors are separated by commas, the last co-author should be separated by "&". In Chinese and Korean names only the surname is generally cited.

Examples:

- *Pipadentalium* Yoo, 1988 (Scaphopoda)
- *Sinentomon* Yin, 1965 (Protura)
- *Belbolla huanghaiensis* Huang & Zhang, 2005 (Nematoda)

Apart from these, there are no commonly accepted conventions. The author can either be spelled following a self-made standard (Linnaeus 1758, Linnaeus 1766), or as given in the original source which implies that names of persons are not always spelled consistently (Linnæus 1758, Linné 1766), or we are dealing with composed data sets without any consistent standard.

Inferred and Anonymous Authorships

In some publications the author responsible for new names and nomenclatural acts is not stated directly in the original source, but can sometimes be inferred from reliable external evidence. Recommendation 51D of the Code states: "...if the authorship is known or inferred from external evidence, the name of the author, if cited, should be enclosed in square brackets to show the original anonymity".

Initials

If the same surname is common to more than one author, initials are sometimes given (for example "A. Agassiz" vs. "L. Agassiz", etc.), but there are no standards concerning this procedure, and not all animal groups / databases use this convention. Although initials are often regarded as useful to disambiguate different persons with the same surname, this does not work in all situations (for example "W. Smith", "C. Pfeiffer", "G. B. Sowerby" and other names occur more than once), and it should be noted that in the examples given in the Code and also the ICZN Official Lists and Indexes, initials are not used.

Implications for Information Retrieval

For a computer, O. F. Müller, O. Müller and Müller are different strings, even the differences between O. F. Müller, O.F. Müller and OF Müller can be problematic. Fauna Europaea is a typical example of a database where combined initials O.F. and O. F. are read as entirely different strings—those who try to search for all taxonomic names described by Otto Friedrich Müller have to know (1) that the submitted data by the various data providers contained several versions (O. F. Müller, O.F. Müller, Müller and O. Müller), and (2) that in many databases, the search function will not find O.F. Müller if you search for O. F. Müller or Müller, not to mention alternative orthographies of this name such as Mueller or Muller.

Thus, the usage of (e.g.) genus-species-author-year, genus-author-year, family-author-year, etc. as "de facto" unique identifiers for biodiversity informatics purposes can present problems, on account of variation in cited author surnames, presence/absence/ variations in cited initials, and minor variants in style of presentation, as well as variant cited authors (responsible person/s) and sometimes, cited dates for what may be in fact the same nomenclatural act in the same work. In addition, in a small number of cases, the same author may have created the same name more than once in the same year for different taxa, which can then only be distinguished by reference to the title, page and sometimes line of the work in which each name appears.

In Australia a program was created (TAXAMATCH) that provides a helpful tool to indicate in a preliminary manner whether two variants of a taxon name should be accepted as identical or not, according to the similarity of the cited author strings. The authority matching function of TAXAMATCH is useful to assign a moderate-to-high similarity to author strings with minor orthographic and/or date differences, such as "Medvedev & Chernov, 1969" vs. "Medvedev & Cernov, 1969", or "Schaufuss, 1877" vs. "L. W. Schaufuss, 1877", or even "Oshmarin, 1952" vs. "Oschmarin in Skrjabin & Evranova, 1952", and a low similarity to author citations which are very different (for example "Hyalesthes Amyot, 1847" vs. "Hyalesthes Signoret, 1865") and are more likely to represent different publication instances, and therefore possibly also different taxa. The program also understands standardized abbreviations as used in Botany and sometimes in Zoology as well, for example "Rchb." for Reichenbach, however may still fail for non-standard abbreviations (such as "H. & A. Ad." for H. & A. Adams, where the normal citation would in fact be "Adams & Adams"); such non-standard abbreviations must then be picked up by subsequent manual inspection after the use of algorithmic approach to pre-sort the names to be matched into groups of either more or less similar names and cited authorities. However, author names which are spelled very similarly but in fact represent different persons, and who independently authored identical taxon names, will not be adequately separated by this program; examples include "O. F. Müller 1776" vs. "P. L. S. Müller 1776", "G. B. Sowerby I 1850" vs. "G. B. Sowerby III 1875" and "L. Pfeiffer 1856" vs. "K. L. Pfeiffer 1956", so additional manual inspection is also required, especially for known problem cases such as those given above.

A further cause of errors that would not be detected by such a program include authors with mult-part surnames which are sometimes inconsistently applied in the literature, and works where the accepted attribution has changed over time. For example, genera published in the anonymously authored work "*Museum Boltenianum sive catalogus cimeliorum...*" published in 1798 were for a long time ascribed to Bolten, but are now considered to have been authored by Röding according to a ruling by the ICZN in 1956. Analogous problems are encountered in the field of attempting to cross-link medical records by patient name, for relevant discussion see record linkage.

Author of a Nomen Nudum

A new name mentioned without description or indication or figure is a *nomen nudum*. A *nomen nudum* has no authorship and date, it is not an available name. If it is desired or necessary to cite the author of such an unavailable name, the nomenclatural status of the name should be made evident.

Sensu Names

A "sensu" name (sensu = "in the sense of", should not be written in italics) is a previously established name that was used by an author in an incorrect sense, for example for a species that was misidentified. Technically this is only a subsequent use of a name, not a new name, and it has no own authorship. Taxonomists often created unwritten rules for authorships of sensu names, to record the first and original source for a misidentification of an animal. But this is not in accordance with the Code.

Example:

- For a West Alpine snail *Pupa ferrari* Porro, 1838, Hartmann (1841) used the genus *Sphyradium* Charpentier, 1837, which Charpentier had established for some similar species. Westerlund argued in 1887 that this species should be placed in another genus, and proposed the name *Coryna* for *Pupa ferrari* and some other species. Pilsbry argued in 1922, Westerlund had established *Coryna* as a new replacement name for *Sphyradium* sensu Hartmann, 1841 (therefore "sensu" should not be written in italics, the term *Sphyradium sensu* Hartmann, 1841 would be misunderstood as a species name). But since a sensu name is not an available name with its own author and year, Pilsbry's argument is not consistent with the ICZN Code's rules.

PHENETICS

In phenetics, also known as taximetrics, organisms are classified based on overall similarity, regardless of their phylogeny or evolutionary relationships. It results in a measure of evolutionary "distance" between taxa. Phenetic methods have become relatively rare in modern times, largely superseded by cladistic analyses, as phenetic methods do not distinguish plesiomorphic from apomorphic traits. However, certain phenetic methods, such as neighbor joining, have found their way into cladistics, as a reasonable approximation of phylogeny when more advanced methods (such as Bayesian inference) are too computationally expensive.

Phenetic techniques include various forms of clustering and ordination. These are sophisticated ways of reducing the variation displayed by organisms to a manageable level. In practice this means measuring dozens of variables, and then presenting them as two- or three-dimensional graphs. Much of the technical challenge in phenetics revolves around balancing the loss of information in such a reduction against the ease of interpreting the resulting graphs.

The method can be traced back to 1763 and Adanson (in his *Familles des plantes*) because of two shared basic principles — overall similarity and equal weighting — and modern pheneticists are sometimes called neo-Adansonians.

Difference from Cladistics

Phenetic analyses are unrooted, that is, they do not distinguish between plesiomorphies, traits that are inherited from an ancestor, and apomorphies, traits that evolved anew in one or several lineages. A common problem with phenetic analysis is that basal evolutionary grades, which retain many plesiomorphies compared to more advanced lineages, appear to be monophyletic. Phenetic analyses are also liable to be misled by convergent evolution and adaptive radiation. Cladistic methods have attempted to solve those problems.

Consider for example songbirds. These can be divided into two groups - Corvida, which retains ancient characters in phenotype and genotype, and Passerida, which has more modern traits. But only the latter are a group of closest relatives; the former are numerous independent and ancient lineages which are about as distantly related to each other as each single one of them is to the Passerida. In a phenetic analysis, the large degree of overall similarity found among the Corvida will make them appear to be monophyletic too, but their shared traits were present in the ancestors of *all* songbirds already. It is the loss of these ancestral traits rather than their presence that signifies which songbirds are more closely related to each other than to other songbirds. However, the requirement that taxa be monophyletic - rather than paraphyletic as in the case of the Corvida - is itself part of the cladistic view of Taxonomy, not necessarily followed to an absolute degree by other schools.

The two methodologies are not mutually exclusive. There is no reason why, e.g., species identified using phenetics cannot subsequently be subjected to cladistic analysis, to determine their evolutionary relationships. Phenetic methods can also be superior to cladistics when only the *distinctness* of related taxa is important, as the computational requirements are lower.

The history of pheneticism and cladism as rival taxonomic systems is analysed in David Hull's 1988 book *Science as a Process*.

Phenetics Today

Traditionally there was a great deal of heated debate between pheneticists and cladists, as both methods were initially proposed to resolve evolutionary relationships. Perhaps the "high-water mark" of phenetics were the DNA-DNA hybridization studies by Charles G. Sibley, Jon E. Ahlquist and Burt L. Monroe Jr., from which resulted the 1990 Sibley-Ahlquist taxonomy for birds. Highly controversial at its time, some of its findings (e.g. the Galloanserae) have been vindicated, while others (e.g. the all-inclusive "Ciconiiformes" or the "Corvida") have been rejected. However, with computers growing increasingly powerful and widespread, more refined cladistic algorithms became available and could put the suggestions of Willi Hennig to the test; as it turned out, the results of cladistic analyses turned out to be superior to those of phenetic methods - at least when it came to resolving phylogenies.

Many systematists continue to use phenetic methods, particularly in addressing species-level questions. While a major goal of taxonomy remains describing the ‹tree of life' - the evolutionary path connecting all species - in fieldwork one needs to be able to separate one taxon from another. Classifying diverse groups of closely related organisms that differ very subtly is difficult using a cladistic approach. Phenetics provides numerical tools for examining overall patterns of variation, allowing researchers to identify discrete groups that can be classified as species.

Modern applications of phenetics are common in botany, and some examples can be found in most issues of the journal *Systematic Botany*. Indeed, due to the effects of horizontal gene transfer, polyploid complexes and other peculiarities of plant genomics, phenetic techniques in botany - though less informative altogether - may, in these special cases, be less prone to errors compared with cladistic analysis of DNA sequences.

In addition, many of the techniques developed by phenetic taxonomists have been adopted and extended by community ecologists, due to a similar need to deal with large amounts of data.

Databases

Modern taxonomy uses database technologies to search and catalogue classifications and their documentation. While there is no commonly used database, there are comprehensive databases such as the *Catalogue of Life*, which attempts to list every documented species. The catalogue listed 1.64 million species for all kingdoms as of April 2016, claiming coverage of more than three quarters of the estimated species known to modern science.

REFERENCES

1. Harper, Douglas. "Taxonomy". Online Etymology Dictionary. Retrieved 21 August 2016.
2. Judd, W.S., Campbell, C.S., Kellogg, E.A., Stevens, P.F., Donoghue, M.J. (2007) Taxonomy. In Plant Systematics – A Phylogenetic Approach, Third Edition. Sinauer Associates, Sunderland.
3. Simpson, Michael G. (2010). "Chapter 1 Plant Systematics: an Overview". Plant Systematics (2nd ed.). Academic Press. ISBN 978-0-12-374380-0.
4. Ruggiero, M.A.; Gordon, D.P.; Orrell, T.M.; Bailly, N.; Bourgoin, T.; Brusca, R.C.; et al. (2015). "A higher level classification of all living organisms". PLoS ONE. 10 (4): e0119248. doi:10.1371/journal.pone.0119248. PMC 4418965 . PMID 25923521.
5. Wilkins, J. S. (5 February 2011). "What is systematics and what is taxonomy?". Retrieved 21 August 2016.
6. Kirk, P.M., Cannon, P.F., Minter, D.W., Stalpers, J.A. eds. (2008) "Taxonomy". In Dictionary of the Fungi, 10th edition. CABI, Netherlands.
7. Walker, P.M.B., ed. (1988). The Wordsworth Dictionary of Science and Technology. W. R. Chambers Ltd. and Cambridge University Press.
8. Lawrence, E. (2005). Henderson's Dictionary Of Biology. Pearson/Prentice Hall. ISBN 9780131273849.
9. Wheeler, Quentin D. (2004). "Taxonomic triage and the poverty of phylogeny". In H. C. J. Godfray & S. Knapp. Taxonomy for the twenty–first century. Philosophical Transactions of the Royal Society. 359. pp. 571–583. doi:10.1098/rstb.2003.1452. PMC 1693342 . PMID 15253345.
10. Singh, Gurcharan (2004). Plant systematics: an integrated approach. Science Publishers. ISBN 1578083516. p. 20.
11. Turrill, W.B. (1938). "The Expansion Of Taxonomy With Special Reference To Spermatophyta". Biological Reviews. 13 (4): 342–373. doi:10.1111/j.1469-185X.1938.tb00522.x.
12. Steyskal, G.C. (1965). "Trend curves of the rate of species description in zoology". Science. 149 (3686): 880–882. Bibcode:1965Sci...149..880S. doi:10.1126/science.149.3686.880. PMID 17737388.
13. Ernst Mayr (1968), "The Role of Systematics in Biology: The study of all aspects of the diversity of life is one of the most important concerns in biology", Science, 159 (3815): 595–599, Bibcode:1968Sci...159..595M, doi:10.1126/science.159.3815.595, PMID 4886900

14. Mayr, E. (1982). "Chapter 6:Microtaxonomy, the science of species". The Growth of Biological Thought: Diversity, Evolution, and Inheritance. Belknap Press of Harvard University Press. ISBN 9780674364462.
15. Datta 1988.
16. Stace 1989.
17. Stuessy 2009.
18. Knapp, S. (2010). "What's in a name? A history of taxonomy".
19. Manktelow, M. (2010) History of Taxonomy. Lecture from Dept. of Systematic Biology, Uppsala University.
20. Bannanje, Govindacharya (2011). Essence of Bhagavata. Udupi: Ishavasya Prathisthana. pp. 57, 58. ISBN 978-81-920104-3-4.
21. Mayr, E. (1982) The Growth of Biological Thought. Belknap P. of Harvard U.P, Cambridge (Mass.)
22. Leroi, Armand Marie (2014). The Lagoon: How Aristotle Invented Science. Bloomsbury. pp. 384–395. ISBN 978-1-4088-3622-4.
23. Linnaeus, C. (1735) Systema naturae, sive regna tria naturae systematice proposita per classes, ordines, genera, & species. Haak, Leiden
24. Linnaeus, C. (1753) Species Plantarum. Stockholm, Sweden.
25. Linnaeus, C. (1758) Systema naturae, sive regna tria naturae systematice proposita per classes, ordines, genera, & species, 10th Edition. Haak, Leiden
26. Donk, M.A. (1957). "Typification and later starting-points" (PDF). Taxon. 6 (9): 245–256. doi:10.2307/1217493. JSTOR 1217493.
27. Secord, James A. (2000). "Victorian Scnsation: Thc Extraordinary Publication, Reception, and Secret Authorship of Vestiges of the Natural History of Creation". Chicago: University of Chicago Press. ISBN 978-0-226-74410-0. Archived from the original on 16 May 2008.
28. Huxley, T. H. (1876): Lectures on Evolution. New York Tribune. Extra. no 36. In Collected Essays IV: pp 46–138 original text w/ figures
29. Rudwick, M. J. S. (1985). The Meaning of Fossils: Episodes in the History of Palaeontology. University of Chicago Press. p. 24. ISBN 0-226-73103-0.
30. Cracraft, Joel and Donaghue, Michael J. (eds.) (2004). Assembling the Tree of Life. Oxford, England: Oxford University Press. ISBN 0195172345. pp. 45, 78 and 555
31. Cavalier-Smith, T. (1998). "A revised six-kingdom system of life". Biological Reviews. 73 (03): 203–66. doi:10.1111/j.1469-185X.1998.tb00030.x. PMID 9809012.
32. Luketa S. (2012). "New views on the megaclassification of life"(PDF). Protistology. 7 (4): 218–237.
33. Linnaeus, C. (1735). Systemae Naturae, sive regna tria naturae, systematics proposita per classes, ordines, genera & species.

34. Haeckel, E. (1866). Generelle Morphologie der Organismen. Reimer, Berlin.
35. Chatton, É. (1925). "Pansporella perplexa. Réflexions sur la biologie et la phylogénie des protozoaires". Annales des Sciences Naturelles - Zoologie et Biologie Animale. 10-VII: 1–84.
36. Copeland, H. (1938). "The kingdoms of organisms". Quarterly Review of Biology. 13: 383–420. doi:10.1086/394568.
37. Whittaker, R. H. (January 1969). "New concepts of kingdoms of organisms". Science. 163 (3863): 150–60. Bibcode:1969Sci...163..150W. doi:10.1126/science.163.3863.150. PMID 5762760.
38. Woese, C.; Kandler, O.; Wheelis, M. (1990). "Towards a natural system of organisms: proposal for the domains Archaea, Bacteria, and Eucarya.". Proceedings of the National Academy of Sciences of the United States of America. 87 (12): 4576–9. Bibcode:1990PNAS...87.4576W. doi:10.1073/pnas.87.12.4576. PMC 54159 . PMID 2112744.
39. Adl, S.M.; Simpson, A.G.B.; Lane, C.E.; Lukeš, J.; Bass, D.; Bowser, S.S.; et al. (2015). "The revised classification of eukaryotes". Journal of Eukaryotic Microbiology. 59: 429–493. doi:10.1111/j.1550-7408.2012.00644.x. PMC 3483872 . PMID 23020233.
40. "What is taxonomy?". Natural History Museum London.
41. "How can I describe new species?". International Commission on Zoological Nomenclature.
42. "About the Catalogue of Life: 2016 Annual Checklist". Catalogue of Life. Integrated Taxonomic Information System (ITIS). Retrieved 22 May 2016.
43. International Code of Zoological Nomenclature, Fourth Edition, 1999 (online version)
44. Blanchard, R., Maehrenthal, F. von & Stiles, C. W. 1905. Règles internationales de la Nomenclature Zoologique adoptées par les Congrès Internationaux de Zoologie. International Rules of Zoological Nomenclature. Internationale Regeln der Zoologischen Nomenklatur. Paris (Rudeval)
45. Sabrosky, C. W. 1974. Article 50 and questions of authorship. Z.N.(S.) 1925. Bulletin of Zoological Nomenclature 31 (4): 206-208.
46. ICZN Code Appendix B 12
47. ICZN Code Recommendation 51B
48. FishBase
49. AnimalBase
50. Fauna Europaea
51. ICZN Official Lists and Indexes (online versions)
52. http://www.faunaeur.org
53. http://www.cmar.csiro.au/datacentre/taxamatch-tests.htm
54. Record for Röding, 1798 reference in AnimalBase

55. ICZN Code Recommendation 51F
56. ICZN Code Art. 51.2.1

CHAPTER 3

Invertebrate Zoology

Chapter Contents

- **Introduction**
- **Subdivisions**
- **Entomology**
- **Carcinology**
- **Subfields**
- **Crayfish**
- **Arachnology**

Invertebrate zoology (IZ) is the study of animal biodiversity. "Invertebrates" include all animals exclusive of vertebrates, in an artificial division set up centuries ago. Of the 34 or so major divisions (phyla) of animal life, 33 and 2/3 are "invertebrate". Animals come in a bewildering diversity of forms and pursue an amazing breadth of life styles, including:

Sponges live attached to the ocean floor, lack a nervous system, and filter water for food. Except some deep sea species turned to snagging plankton with specialized, hooked spicules.

Corals build reefs visible from space, the largest structures created by living things, including humans. Internal microscopic algae feed them, but they also indulge in carnivorous fare. Corals host countless other species in their complex colonies.

Worms... The word may repulse some, but for a zoologist, it means diversity. Over half of the animal phyla are "worms" - elongate, muscular organisms that swim in the ocean, burrow in the sea floor or reef, live in soil, or crawl inside most plants and animals large enough to host "helminth" parasites. Feather duster worms, pin worms, large flat worms that swim like flying carpets are unrelated "invertebrates" all sporting a "wormy" body plan.

Mollusks, the second largest phylum, are big not only in species diversity but in body form: worms, clams, slugs, flying squid, tusk shells, chitons, limpets, sea butterflies.

FLMNH hosts about a third of the 100,000 known species. More species of arthropods are known than of all other animals put together. A better division of the world would be into Arthropods and Unarthropods... Body forms range from parasitic, pulsating blobs, to centipedes, horseshoe crabs, and butterflies.

Echinoderms include the conspicuous sea stars, sea cucumbers, and sea urchins. They are our symmetrical, large, striking, but weird cousins. How many of your other cousins can change the consistency of their body to turn to slime, or to lock their arm in a position so that it will break before unbending?

Chordates are the most familiar animals, as they include fish, frogs, eagles, and humans. But they also include lancelets, salps, larvaceans, and sea squirts. Animals that make mucus houses several times their own size for filter feeding. Bioluminescent colonies 30 feet long that jet propel through the ocean. Brainless, fish-like lancelets that burrow in the ocean floor to filter-feed on plankton.

INTRODUCTION

Invertebrate zoology is the biological discipline that consists of the study of invertebrate animals, i.e. animals without a backbone (a structure which is found only in fish, amphibians, reptiles, birds and mammals.)

Invertebrates are a vast and very diverse group of animals that includes sponges, echinoderms, tunicates, numerous different phyla of worms, molluscs, arthropods and many additional phyla. Single-celled organisms or protists are usually not included within the same group as invertebrates.

SUBDIVISIONS

Invertebrates are 97% of all named animal species, and because of that fact, this subdivision of zoology has many further subdivisions, including but not limited to:

- Arthropodology - the study of arthropods, which includes
- Arachnology - the study of spiders and other arachnids
- Entomology - the study of insects
- Carcinology - the study of crustaceans
- Myriapodology - the study of centipedes, millipedes, and other myriapods
- Cnidariology - the study of Cnidaria
- Helminthology - the study of parasitic worms.
- Malacology - the study of mollusks, which includes
- Conchology - the study of Mollusk shells.
- Limacology - the study of slugs.
- Teuthology - the study of cephalopods.
- Invertebrate paleontology - the study of fossil invertebrates

These divisions are sometimes further divided into more specific specialties. For example, within arachnology, acarology is the study of mites and ticks; within entomology, lepidoptery is the study of butterflies and moths, Myrmecology is the study of ants and so on.

ENTOMOLOGY

A phasmid, mimicking a leaf

Entomology (from Greek ντομον, entomon "insect"; and -λογία, -logia) is the scientific study of insects, a branch of zoology. In the past the term "insect" was more vague, and historically the definition of entomology included the study of terrestrial animals in other arthropod groups or other phyla, such as arachnids, myriapods, earthworms, land snails, and slugs. This wider meaning may still be encountered in informal use.

Like several of the other fields that are categorized within zoology, entomology is a taxon-based category; any form of scientific study in which there is a focus on insect-related inquiries is, by definition, entomology. Entomology therefore overlaps with a cross-section of topics as diverse as molecular genetics, behavior, biomechanics, biochemistry, systematics, physiology, developmental biology, ecology, morphology, paleontology

At some 1.3 million described species, insects account for more than two-thirds of all known organisms, date back some 400 million years, and have many kinds of interactions with humans and other forms of life on earth.

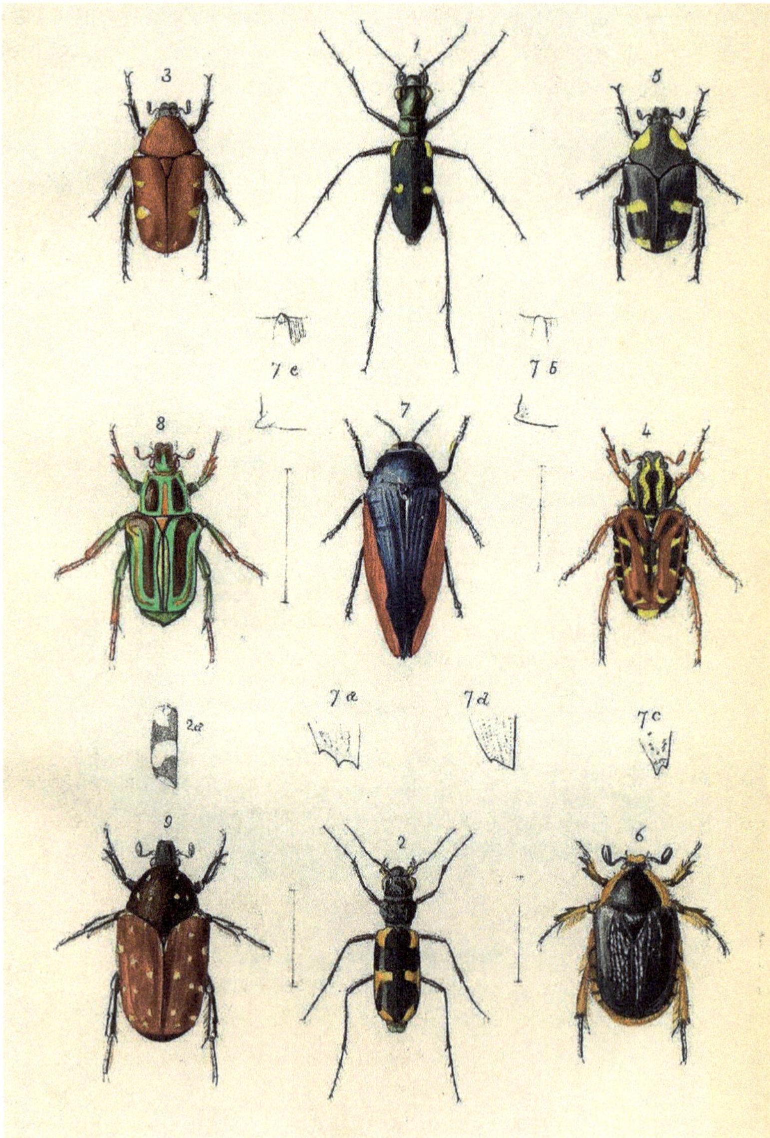

Plate from Transactions of the Entomological Society, 1848

Entomology is rooted in nearly all human cultures from prehistoric times, primarily in the context of agriculture (especially biological control and beekeeping), but scientific study began only as recently as the 16th century.

William Kirby is widely considered as the father of Entomology. In collaboration with William Spence, he published a definitive entomological encyclopedia, Introduction to Entomology, regarded as the subject's foundational text. He also helped to found the Royal Entomological Society in London in 1833, one of the earliest such societies in the world; earlier antecedents, such as the Aurelian society date back to the 1740s.

Entomology developed rapidly in the 19th and 20th centuries, and was studied by large numbers of people, including such notable figures as Charles Darwin, Jean-Henri

Fabre, Vladimir Nabokov, Karl von Frisch (winner of the 1973 Nobel Prize in Physiology or Medicine), and two-time Pulitzer Prize winner E. O. Wilson.

These 100 Trigonopterus species were described simultaneously using DNA barcoding

Most insects can easily be recognized to order such as Hymenoptera (bees, wasps, and ants) or Coleoptera (beetles). However, insects other than Lepidoptera (butterflies and moths) are typically identifiable to genus or species only through the use of Identification keys and Monographs. Because the class Insecta contains a very large number of species (over 330,000 species of beetles alone) and the characteristics separating them are unfamiliar, and often subtle (or invisible without a microscope), this is often very difficult even for a specialist. This has led to the development of automated species identification systems targeted on insects, for example, Daisy, ABIS, SPIDA and Draw-wing

Insect identification is an increasingly common hobby, with butterflies and dragonflies being the most popular.

In Pest Control

In 1994 the Entomological Society of America launched a new professional certification program for the pest control industry called The Associate Certified Entomologist (ACE). To qualify as a "true entomologist" an individual would normally require an advanced degree, with most entomologists pursuing their PhD. While not true entomologists in the traditional sense, individuals who attain the ACE certification may be referred to as ACEs, Amateur entomologists, Associate entomologists or –more commonly– Associate-Certified Entomologists.

Taxonomic Specialization

Part of a large beetle collection

Many entomologists specialize in a single order or even a family of insects, and a number of these subspecialties are given their own informal names, typically (but not always) derived from the scientific name of the group:

- Coleopterology - beetles
- Dipterology - flies
- Hemipterology - true bugs
- Isopterology - termites
- Lepidopterology - moths and butterflies
- Melittology (or Apiology) - bees

- Myrmecology - ants
- Orthopterology - grasshoppers, crickets, etc.
- Trichopterology - caddis flies
- Vespology - Social wasps

Organizations

Like other scientific specialties, entomologists have a number of local, national, and international organizations. There are also many organizations specializing in specific subareas.

- Amateur Entomologists' Society
- Deutsches Entomologisches Institut
- Entomological Society of America
- Entomological Society of Canada
- Entomological Society of Japan
- International Union for the Study of Social Insects
- Netherlands Entomological Society
- Royal Belgian Entomological Society
- Royal Entomological Society of London
- Société entomologique de France

Museums

Here is a list of selected museums which contain very large insect collections.

Africa

- Natal Museum, Pietermaritzburg, South Africa

Europe

- Muséum national d'histoire naturelle, Paris, France
- Museum für Naturkunde, Berlin, Germany
- Natural History Museum, Budapest Hungarian Natural History Museum
- Natural History Museum, Geneva
- Natural History Museum, Leiden, the Netherlands
- Natural History Museum, London, United Kingdom
- Natural History Museum, Oslo Norway
- Natural History Museum, St. Petersburg Zoological Collection of the Russian

Academy of Science
- Naturhistorisches Museum, Vienna, Austria
- Oxford University Museum of Natural History, Oxford
- Royal Museum for Central Africa, Brussels, Belgium
- Swedish Museum of Natural History, Stockholm, Sweden
- The Bavarian State Collection of Zoology Zoologische Staatssammlung München
- World Museum Liverpool, the Bug House

United States

- Academy of Natural Sciences of Philadelphia
- American Museum of Natural History, New York City
- Auburn University Museum of Natural History, Auburn, Alabama
- Audubon Insectarium, New Orleans
- Bohart Museum of Entomology, Davis, California
- California Academy of Sciences, San Francisco
- Carnegie Museum of Natural History, Pittsburgh
- Essig Museum, Berkeley, California
- Field Museum of Natural History, Chicago
- Florida Museum of Natural History, University of Florida, Gainesville, Florida
- Illinois Natural History Survey, Champaign, Illinois
- J. Gordon Edwards Museum, San Jose, California
- Museum of Comparative Zoology, Cambridge, Massachusetts
- Natural History Museum of Los Angeles County, Los Angeles
- National Museum of Natural History, Washington, D.C.
- New Mexico State University Arthropod Museum
- North Carolina State University Insect Museum, Raleigh, North Carolina
- Peabody Museum of Natural History, New Haven, Connecticut
- The National Museum of Play, Rochester, N.Y.
- Texas A&M University, College Station, Texas
- University of Minnesota, St. Paul campus (UMSP), Minnesota
- University of Kansas Natural History Museum, Lawrence, Kansas
- University of Nebraska State Museum, Lincoln, Nebraska
- University of Missouri Enns Entomology Museum, University of Missouri, Columbia, Missouri

Canada

- Canadian Museum of Nature, Ottawa
- Canadian National Collection of Insects, Arachnids and Nematodes, Ottawa, Ontario
- E.H. Strickland Entomological Museum, University of Alberta, Edmonton, Alberta
- Lyman Entomological Museum, Macdonald Campus of McGill University, Sainte-Anne-de-Bellevue, Quebec
- Montreal Insectarium, Montreal, Quebec
- Newfoundland Insectarium, Reidville, Newfoundland and Labrador
- Royal Alberta Museum, Edmonton, Alberta
- Royal Ontario Museum, Toronto
- University of Guelph Insect Collection, Guelph, Ontario
- Victoria Bug Zoo, Victoria, British Columbia

CARCINOLOGY

Carcinology is a branch of zoology that consists of the study of crustaceans, a group of arthropods that includes lobsters, crayfish, shrimp, krill, barnacles and crabs. Other names for carcinology are malacostracology, crustaceology, and crustalogy, and a person who study crustaceans is a carcinologist or occasionally a malacostracologist, a crustaceologist, or a crustalogist.

The word carcinology derives from Greek καρκίνος, karkínos, "crab"; and -λογία, -logia.

SUBFIELDS

Carcinology is a subdivision of arthropodology, the study of arthropods which includes arachnids, insects, and myriapods. Carcinology branches off into taxonomically oriented disciplines such as:

- astacology – the study of crayfish
- cirripedology – the study of barnacles
- copepodology – the study of copepods

CRAYFISH

Crayfish, also known as crawfish, crawdads, freshwater lobsters, mountain lobsters, mudbugs or yabbies, are freshwater crustaceans resembling small lobsters, to which they are related; taxonomically, they are members of the superfamilies Astacoidea and Parastacoidea. They breathe through feather-like gills. Some species are found in brooks and streams where there is running fresh water, while others thrive in swamps, ditches, and rice paddies. Most crayfish cannot tolerate polluted water, although some species such as Procambarus clarkii are hardier. Crayfish feed on living and dead animals and plants.

Names

The name "crayfish" comes from the Old French word escrevisse (Modern French écrevisse). The word has been modified to "crayfish" by association with "fish" (folk etymology). The largely American variant "crawfish" is similarly derived.

Some kinds of crayfish are known locally as lobsters, crawdads, mudbugs, and yabbies. In the Eastern United States, "crayfish" is more common in the north, while "crawdad" is heard more in central and southwestern regions, and "crawfish" further south, although there are considerable overlaps.

The study of crayfish is called astacology.

Other Animals

In Australia (on the eastern seaboard), New Zealand and South Africa, the term crayfish or cray generally refers to a saltwater spiny lobster, of the genus Jasus that is indigenous to much of southern Oceania, while the freshwater species are usually called yabby or kōura, from the indigenous Australian and Māori names for the animal respectively, or by other names specific to each species. Exceptions include western rock lobster (of the Palinuridae family) found on the west coast of Australia; the Tasmanian giant freshwater crayfish (from the Parastacidae family) found only in Tasmania; and the Murray crayfish found on Australia's Murray River.

In Singapore, the term crayfish typically refers to Thenus orientalis, a seawater crustacean from the slipper lobster family. True crayfish are not native to Singapore, but are commonly found as pets, or as an invasive species (Cherax quadricarinatus) in the many water catchment areas, and are alternatively known as freshwater lobsters.

Anatomy

The body of a decapod crustacean, such as a crab, lobster, or prawn (shrimp), is made up of twenty body segments grouped into two main body parts, the cephalothorax and the abdomen. Each segment may possess one pair of appendages, although in various groups these may be reduced or missing. On average, crayfish grow to 17.5 centimetres (6.9 in) in length, but some grow larger. Walking legs have a small claw at the end.

Geographical Distribution and Classification

The three Families of Crayfish

Astacidae: Austropotamobius pallipes

Cambaridae: Procambarus alleni

Parastacidae: Cherax pulcher.

There are three families of crayfish, two in the Northern Hemisphere and one in the Southern Hemisphere. The Southern Hemisphere (Gondwana-distributed) family Parastacidae lives in South America, Madagascar and Australasia, and is distinguished by the lack of the first pair of pleopods. Of the other two families, members of the Astacidae live in western Eurasia and western North America and members of the family Cambaridae live in eastern Asia and eastern North America.

Madagascar has an endemic genus, Astacoides, containing seven species.

Europe is home to seven species of crayfish in the genera Astacus and Austropotamobius.

Cambaroides is native to Japan and eastern mainland Asia.

North America

The greatest diversity of crayfish species is found in southeastern North America, with over 330 species in nine genera, all in the family Cambaridae. A further genus of astacid crayfish is found in the Pacific Northwest and the headwaters of some rivers east of the Continental Divide. Many crayfish are also found in lowland areas where the water is abundant in calcium, and oxygen rises from underground springs. Crayfish were introduced purposely into a few Arizona reservoirs and other bodies of water decades ago, primarily as a food source for sport fish. They have since dispersed beyond those original sites.

Australia

Australia has over 100 species in a dozen genera. In Australia, many of the better-known crayfish are of the genus Cherax, and include the marron from Western Australia (now believed to be two species, Cherax tenuimanus and C. cainii), red-claw crayfish (Cherax quadricarinatus), common yabby (Cherax destructor) and western yabby (Cherax preissii). The marron are some of the largest crayfish in the world. They grow up to several pounds in size. C. tenuimanus is critically endangered, while other large Australasian crayfish are threatened or endangered. Australia is home to the world's two largest freshwater crayfish – the Tasmanian giant freshwater crayfish Astacopsis gouldi, which can achieve a mass of over 5 kilograms (11 lb) and is found in rivers of northern Tasmania, and the Murray crayfish Euastacus armatus, which can reach 2 kilograms (4.4 lb) and is found in much of the southern Murray-Darling basin.

New Zealand

In New Zealand, two species of Paranephrops are endemic, and are known by the Māori name kōura.

Fossil record

Fossil records of crayfish older than 30 million years are rare, but fossilised burrows have been found from strata as old as the late Palaeozoic or early Mesozoic. The oldest records of the Parastacidae are in Australia, and are 115 million years old.

Crayfish Plague

Some crayfish suffer from a disease called crayfish plague, caused by the North American water mould Aphanomyces astaci which was transmitted to Europe when North American species of crayfish were introduced there. Species of the genus Astacus are particularly susceptible to infection, allowing the plague-coevolved signal crayfish to invade parts of Europe.

Uses

Food

Crayfish, boiled

Crayfish are eaten worldwide. Like other edible crustaceans, only a small portion of the body of a crayfish is eaten. In most prepared dishes, such as soups, bisques and étouffées, only the tail portion is served. At crawfish boils or other meals where the entire body of the crayfish is presented, other portions, such as the claw meat, may be eaten. Like all crustaceans, crayfish are not kosher because they are aquatic animals that do not have both fins and scales. They are therefore not eaten by observant Jews. As of 2005, Louisiana supplies 95% of the crayfish harvested in the US. In 1987, Louisiana produced 90% of the crayfish harvested in the world, 70% of which were consumed locally. In 2007, the Louisiana crawfish harvest was about 54,800 tons, almost all of it from aquaculture. About 70%–80% of crayfish produced in Louisiana are Procambarus clarkii (red swamp crawfish), with the remaining 20%–30% being Procambarus zonangulus (white river crawfish).

Bait

Crayfish are commonly sold and used as bait, either live or with only the tail meat, and are good at attracting channel catfish, walleye, trout, largemouth bass, smallmouth bass, pike and muskellunge. Sometimes the claws are removed so that the crayfish do not stop fish from biting the hook. Crayfish easily fall off the hook, so casting should be slow.

The result of using crayfish as bait has led to various ecological problems at times. According to a report prepared by Illinois State University, on the Fox River and Des Plaines River watershed, "The rusty crayfish (used as bait) has been dumped into the water and its survivors outcompete the native clearwater crayfish". This situation has been repeated elsewhere, as the crayfish bait eliminates native species.

The use of crayfish as bait has been cited as one of the ways zebra mussels have spread to different waterways, as members of this invasive species are known to attach themselves to crayfish.

Pets

Procambarus clarkii in a freshwater aquarium

Crayfish are kept as pets in freshwater aquariums. Crayfish kept as pets in the US from local waters are usually kept with bluegill or bass, rather than goldfish or tropical or subtropical fish. They prefer foods like shrimp pellets or various vegetables, but will also eat tropical fish food, regular fish food, algae wafers, and small fish that can be captured with their claws. They will sometimes consume their old exoskeleton after it has moulted. Their disposition towards eating almost anything will also cause them to explore the edibility of aquarium plants in a fish tank. However, most species of dwarf crayfish, such as Cambarellus patzcuarensis, will not destructively dig or eat live aquarium plants. They are also relatively non-aggressive and can be kept safely with dwarf shrimp. Because of their very small size of 1.5 inches (38 mm) or less, some fish are often a threat to the crayfish.

Since crayfish are accustomed to being in ponds or rivers, they will have a tendency to shift gravel around on the bottom of the tank, creating mounds or trenches to emulate a burrow. Crayfish will often try to climb out of the tank, especially if an opening exists at the top that they can fit through.

In some nations, such as the United Kingdom, United States, Australia, and New Zealand, imported alien crayfish are a danger to local rivers. The three species commonly imported to Europe from the Americas are Orconectes limosus, Pacifastacus leniusculus and Procambarus clarkii.Crayfish may spread into different bodies of water because specimens captured for pets in one river are often released into a different catchment. There is a potential for ecological damage when crayfish are introduced into non-native bodies of water (e.g., crayfish plague in Europe).

ARACHNOLOGY

Nephila clavipes

Arachnology is the scientific study of spiders and related animals such as scorpions, pseudoscorpions, and harvestmen, collectively called arachnids. Those who study spiders and other arachnids are arachnologists. The word arachnology derives from Greek ῥάχνη, arachnē, "spider"; and -λογία, -logia.

Arachnology as a Science

Arachnologists are primarily responsible for classifying arachnids and studying aspects of their biology. In the popular imagination they are sometimes referred to as 'spider experts'. Disciplines within arachnology include naming species and determining their evolutionary relationships to one another (taxonomy and systematics), studying how they interact with other members of their species and/or their environment (behavioural ecology), or how they are distributed in different regions and habitats (faunistics). Other arachnologists perform research on the anatomy or physiology of arachnids, including the venom of spiders and scorpions. Others study the impact of spiders in agricultural ecosystems and whether they can be used as biological control agents.

Subdisciplines

Arachnology can be broken down into more specific specialties. These topics include:

- acarology - the study of ticks and mites.
- araneology - the study of spiders.
- scorpiology - the study of scorpions.

Arachnological societies

Arachnologists are served by a number of scientific societies, both national and international in scope. Their main role is to encourage the exchange of ideas between researchers, to organise meetings and congresses, and in a number of cases to publish academic journals. Some are also involved in outreach programs, like the European spider of the year, which raise awareness of these animals among the general public.

- American Arachnological Society (AAS)
- British Arachnological Society (BAS)
- International Society of Arachnology (ISA)
- Iranian Arachnological Society (IAS)
- African Arachnological Society (AFRAS)
- Arachnologia Belgica - Belgian Arachnological Society (ARABEL)
- The Arachnological Society of Japan (ASJ)
- Arachnologische Gesellschaft - Central European Arachnological Society
- Asian Society of Arachnology (ASA)
- L'Association Francaise d'Arachnologie - French Society of Arachnology (AsFrA)
- The Australasian Arachnological Society
- Czech Arachnological Society

- Deutsche Arachnologische Gesellschaft e. V. - German Arachnologic Society (DeArGe)
- European Society of Arachnology (ESA)
- Grupo Ibérico de Aracnologia - Spanish Society of Arachnology (Iberian Peninsula)
- Indian Society of Arachnology
- Magyar Arachnolgia - Hungarian Arachnology

Arachnological Journals

Scientific journals devoted to the study of arachnids include:

- Bulletin of the British Arachnological Society
- Journal of Arachnology

Popular Arachnology

In the 1970s, arachnids - particularly tarantulas - started to become popular as exotic pets. Many tarantulas thus become more widely known by their common names such as the Mexican redknee tarantula (Brachypelma smithi).

Various societies now focus on the husbandry, care, study and captive breeding of tarantulas, and other arachnids. They also typically produce journals or newsletters with articles and advice on these subjects.

- British Tarantula Society
- The American Tarantula Society

REFERNCES

1. Christoph Needon; Johannes Petermann; Peter Scheffel; Bernd Scheibe (1971). Plants and Animals (Pflanzen und Tiere). Leipzig: Urania Verlag.
2. "crayfish". Oxford English Dictionary (3rd ed.). Oxford University Press. September 2005. (Subscription or UK public library membership required.)
3. C. W. Hart, Jr. (1994). "A dictionary of non-scientific names of freshwater crayfishes (Astacoidea and Parastacoidea), including other words and phrases incorporating crayfish names". Smithsonian Contributions to Anthropology. 38 (38): 1–127. doi:10.5479/si.00810223.38.1. hdl:10088/1372.
4. ^ Jump up to:[a] [b] Pableaux Johnson. "Mudbug Madness : Crawfish". Bayou Dog. Archived from the original on 19 August 2006. Retrieved 28 August 2006.
5. Bert Vaux; Scott A. Golder. "Dialect survey". Harvard University. Retrieved 30 September 2006.
6. «About the International Association of Astacology". Archived from the original on 5 April 2005.
7. Harold W. Sims Jr. (1965). "Let's call the spiny lobster "spiny lobster"".

Crustaceana. 8 (1): 109–110. doi:10.1163/156854065X00613. JSTOR 20102626.

8. «Sweet Chilli Crayfish “. mywoklife.com. 13 February 2010.
9. «FAR OCEAN SEA PRODUCTS (PRIVATE) LIMITED”. dollarvietnam.com.
10. Classic Asian Noodles. Marshall Cavendish. 2007. ISBN 9812613358.
11. Shane T. Ahyong; Darren C. J. Yeo (2007). “Feral populations of the Australian Red-Claw crayfish (Cherax quadricarinatus von Martens) in water supply catchments of Singapore”. Biol Invasions. 9 (8): 943–946. doi:10.1007/s10530-007-9094-0.
12. «Crayfish (aka Freshwater Lobster) – Arofanatics Fish Talk Forums”. 9 December 2004.
13. Horton H. Hobbs, Jr. (1974). “Synopsis of the families and genera of crayfishes (Crustacea: Decapoda)” (PDF). Smithsonian Contributions to Zoology. 164 (164): 1–32. doi:10.5479/si.00810282.164.
14. Christopher B. Boyko; Olga Ramilijaona Ravoahangimalala; Désiré Randriamasimanana; Tony Harilala Razafindrazaka (2005). “Astacoides hobbsi, a new crayfish (Crustacea: Decapoda: Parastacidae) from Madagascar” (PDF). Zootaxa. 1091: 41–51.
15. Steve Pollock (2005). Eyewitness Ecology. New York, United States: Dorling Kindersley. ISBN 978-0-7894-5581-9.
16. Michael P. Masser; David B. Rouse (1997). “Australian Red Claw Crayfish” (PDF). SRAC Publication. Southern Regional Aquaculture Center (244). Archived from the original (PDF) on 13 May 2005.
17. «Tasmanian Giant Freshwater Lobster (Astacopsis gouldi)”. Department of the Environment, Water, Heritage and the Arts. 9 February 2007. Retrieved 16 March 2010.
18. “Kōura”. NIWA. Retrieved 18 September 2012.
19. Alycia L. Rode & Loren E. Babcock (2003). “Phylogeny of fossil and extant freshwater crayfish and some closely related nephropid lobsters”. Journal of Crustacean Biology. 23 (2): 418–435. doi:10.1651/0278-0372(2003)023[0418:POFAEF]2.0.CO;2. JSTOR 1549646.
20. Emory University (12 February 2008). “Oldest Australian crayfish fossils provide missing evolutionary link”. ScienceDaily.
21. Jump up to:[a] [b] James R. Lee (5 December 1998). “TED Case Studies Crayfish Plague #478 European Crayfish Dispute”. Archived from the original on 10 January 2009. Retrieved 20 January 2008.
22. “Kosher defined”. Triangle K. Retrieved 1 December 2010.
23. Robert P. Romaire, W. Ray McClain, Mark G. Shirley and C. Greg Lutz, Crawfish Aquaculture — Marketing (SRAC Publication No. 2402). October 2005. Southern Regional Aquaculture Center.
24. Larry W. de la Bretonne, Jr. & Robert P. Romaire (1990). “Crawfish production: harvesting, marketing and economics” (PDF). SRAC Publication. Southern

Regional Aquaculture Center (242). Archived from the original (PDF) on 12 December 2010.

25. "1978–2007: Louisiana Summary of Agriculture and Natural Resources" (PDF). Louisiana State University Agricultural Center, Baton Rouge, Louisiana. 2009. Archived from the original (PDF) on 2 April 2012.
26. "Differences Between Red Swamp Crawfish and White River Crawfish". The Louisiana State University Agricultural Center.
27. "Fox and Des Plaines Rivers Watershed". Critical Trends in Illinois Ecosystems (PDF). Illinois Department of Natural Resources. 2001.
28. Tennessee Aquatic Nuisance Species Task Force (2007). Tennessee Aquatic Nuisance Species Management Plan (PDF). Tennessee Wildlife Resources Agency.
29. J. Thompson; F. Parchaso; A. Alpine; J. Cloern; B. Cole; O. Mace; J. Edmunds; J. Baylosis; S. Luoma; F. Nichols (13 December 2007). "The History and Effects of Exotic Species in San Francisco Bay". United States Geological Survey.
30. Gerald Pottern. "Mexican dwarf orange crayfish, Cambarellus patzcuarensis". Retrieved 13 October 2010.
31. Flatow, Ira (17 October 2014). "Environmental Detectives Use Genetic Tools to Track Invasives" (Podcast). Public Radio International. Event occurs at 9:14.

CHAPTER 4

Vertebrate Zoology

Chapter Contents

- **Introduction**
- **Ichthyology**
- **Fish**
- **Importance To Humans**
- **Terminology**

Vertebrate Zoology is the study of animals with backbones. Research in the department covers fishes, amphibians, reptiles, birds, and mammals.

INTRODUCTION

Vertebrate zoology is the biological discipline that consists of the study of Vertebrate animals, i.e., animals with a backbone, such as fish, amphibians, reptiles, birds and mammals.

This subdivision of zoology has many further subdivisions, including:

- Ichthyology - the study of fishes.
- Mammalogy - the study of mammals.
- Ornithology - the study of birds.
- Herpetology - the study of reptiles.
- Batrachology - the study of amphibians.

ICHTHYOLOGY

Coelacanth

Ichthyology, also known as **fish science**, is the branch of biology devoted to the study of fish. This includes bony fishes (Osteichthyes), cartilaginous fish (Chondrichthyes), and jawless fish (Agnatha). While a large number of species have been discovered, approximately 250 new species are officially described by science each year. According to FishBase, 33,400 species of fish had been described by October 2016.

Fish represent approximately 8% of all figurative depictions on Mimbres pottery.

The study of fish dates from the Upper Paleolithic Revolution (with the advent of "high culture"). The science of ichthyology was developed in several interconnecting epochs, each with various significant advancements.

The study of fish receives its origins from human's desire to feed, clothe, and equip themselves with useful implements. According to Michael Barton, a prominent ichthyologist and professor at Centre College, "the earliest ichthyologists were hunters and gatherers who had learned how to obtain the most useful fish, where to obtain them in abundance, and at what times they might be the most available". Early cultures manifested these insights in abstract and identifiable artistic expressions.

1500 BC–40 AD

Informal, scientific descriptions of fish are represented within the Judeo-Christian tradition. The Old Testament laws of kashrut forbade the consumption of fishes without scales or appendages. Theologians and ichthyologists speculate that the apostle Peter and his contemporaries harvested the fish that are today sold in modern industry along the Sea of Galilee, presently known as Lake Kinneret. These fish include cyprinids of the genera *Barbus* and *Mirogrex*, cichlids of the genus *Sarotherodon*, and *Mugil cephalus* of the family Mugilidae.

335 BC–80 AD

Aristotle incorporated ichthyology into formal scientific study. Between 335 BC–322 BC, he provided the earliest taxonomic classification of fish, accurately describing 117 species of Mediterranean fish. Furthermore, Aristotle documented anatomical and behavioral differences between fish and marine mammals. After his death, some of his pupils continued his ichthyological research. Theophrastus, for example, composed a treatise on amphibious fish. The Romans, although less devoted to science, wrote extensively about fish. Pliny the Elder, a notable Roman naturalist, compiled the ichthyological works of indigenous Greeks, including verifiable and ambiguous peculiarities such as the sawfish and mermaid respectively. Pliny's documentation was the last significant contribution to ichthyology until the European Renaissance.

European Renaissance

The writings of three sixteenth century scholars, Hippolito Salviani, Pierre Belon, and Guillaume Rondelet, signify the conception of modern ichthyology. The investigations of these individuals were based upon actual research in comparison to ancient recitations. This property popularized and emphasized these discoveries. Despite their prominence, Rondelet's *De Piscibus Marinis* is regarded as the most influential, identifying 244 species of fish.

16th–17th century

The incremental alterations in navigation and shipbuilding throughout the Renaissance marked the commencement of a new epoch in ichthyology. The Renaissance culminated with the era of exploration and colonization, and upon the cosmopolitan interest in navigation came the specialization in naturalism. Georg Marcgrave of Saxony composed the *Naturalis Brasilae* in 1648. This document contained a description of 100 species of fish indigenous to the Brazilian coastline. In 1686, John Ray and Francis Willughby collaboratively published *Historia Piscium*, a scientific manuscript containing 420 species of fish, 178 of these newly discovered. The fish contained within this informative literature were arranged in a provisional system of classification.

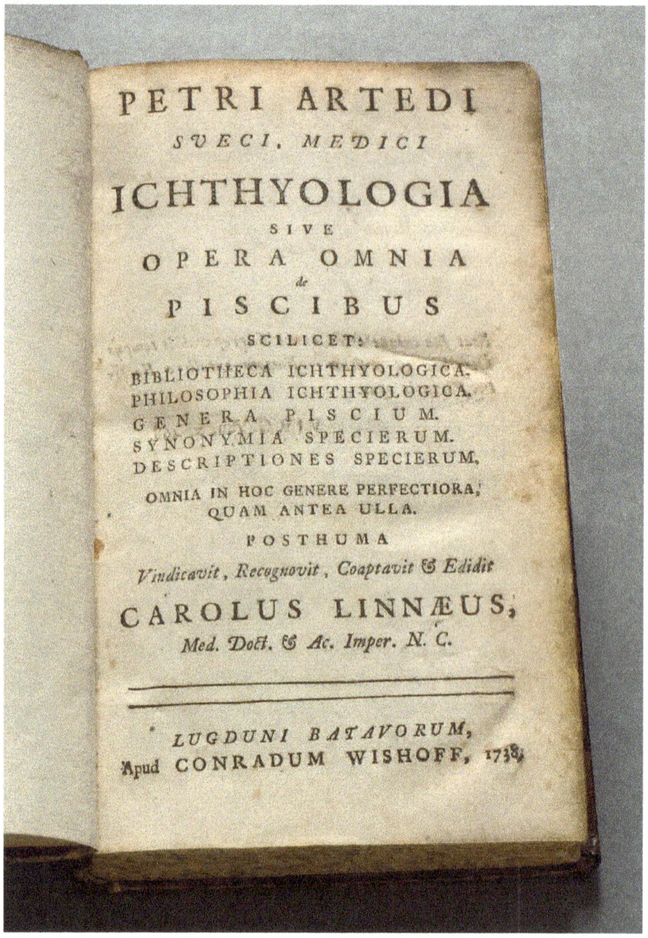
PETRI ARTEDI
SVECI, MEDICI
ICHTHYOLOGIA
SIVE
OPERA OMNIA
de
PISCIBUS
SCILICET:
BIBLIOTHECA ICHTHYOLOGICA.
PHILOSOPHIA ICHTHYOLOGICA.
GENERA PISCIUM.
SYNONYMIA SPECIERUM.
DESCRIPTIONES SPECIERUM.
OMNIA IN HOC GENERE PERFECTIORA,
QUAM ANTEA ULLA.
POSTHUMA
Vindicavit, Recognovit, Coaptavit & Edidit
CAROLUS LINNÆUS,
Med. Doct. & Ac. Imper. N. C.
LUGDUNI BATAVORUM,
Apud CONRADUM WISHOFF, 1738.

Frontipiece from *Ichthyologia, sive Opera Omnia de Piscibus* by Peter Artedi

The classification used within the *Historia Piscium* was further developed by Carl Linnaeus, the "father of modern taxonomy". His taxonomic approach became the systematic approach to the study of organisms, including fish. Linnaeus was a professor at the University of Uppsala and an eminent botanist; however, one of his colleagues, Peter Artedi, earned the title "father of ichthyology" through his indispensable advancements. Artedi contributed to Linnaeus's refinement of the principles of taxonomy. Furthermore, he recognized five additional orders of fish: Malacopterygii, Acanthopterygii, Branchiostegi, Chondropterygii, and Plagiuri. Artedi developed standard methods for making counts and measurements of anatomical features that are modernly exploited. Another associate of Linnaeus, Albertus Seba, was a prosperous pharmacist from Amsterdam. Seba assembled a cabinet, or collection, of fish. He invited Artedi to utilize this assortment of fish; unfortunately, in 1735, Artedi fell into an Amsterdam canal and drowned at the age of 30.

Linnaeus posthumously published Artedi's manuscripts as *Ichthyologia, sive Opera Omnia de Piscibus* (1738). His refinement of taxonomy culminated in the development of the binomial nomenclature which is in use by contemporary ichthyologists. Furthermore, he revised the orders introduced by Artedi, placing significance on pelvic fins. Fishes lacking this appendage were placed within the order Apodes; fishes containing abdominal, thoracic, or jugular pelvic fins were termed Abdominales, Thoracici, and Jugulares respectively. However, these alterations were not grounded within evolutionary theory. Therefore, it would take over a century until Charles Darwin would provide the intellectual foundation from which we would be permitted to perceive that the degree of similarity in taxonomic features was a consequence of phylogenetic relationship.

Modern Era

Close to the dawn of the nineteenth century, Marcus Elieser Bloch of Berlin and Georges Cuvier of Paris made attempts to consolidate the knowledge of ichthyology. Cuvier summarized all of the available information in his monumental *Histoire Naturelle des Poissons*. This manuscript was published between 1828 and 1849 in a 22 volume series. This documental describes 4,514 species of fish, 2,311 of these new to science. It remains one of the most ambitious treatises of the modern world. Scientific exploration of the Americas advanced our knowledge of the remarkable diversity of fishes. Charles Alexandre Lesueur was a student of Cuvier. He made a cabinet of fishes dwelling within the Great Lakes and Saint Lawrence River regions.

Adventurous individuals such as John James Audubon and Constantine Samuel Rafinesque figure in the faunal documentation of North America. These persons often traveled with one another. Rafinesque wrote *Ichthyologia Ohiensis* in 1820. In addition, Louis Agassiz of Switzerland established his reputation through the study of freshwater fishes and the first comprehensive treatment of paleoichthyology, Poissons Fossiles. In the 1840s, Agassiz moved to the United States, where he taught at Harvard University until his death in 1873.

Albert Günther published his *Catalogue of the Fishes of the British Museum* between 1859 and 1870, describing over 6,800 species and mentioning another 1,700. Generally considered one of the most influential ichthyologists, David Starr Jordan wrote 650 articles and books on the subject as well as serving as president of Indiana University and Stanford University.

Modern Publications

Publication	**Frequency**	**Date of publication**	**Publisher**
Copeia	Quarterly	27 December 1913	American Society of Ichthyologists and Herpetologists
Journal of Applied Ichthyology	Bi-monthly	1985	Blackwell Publishing
Ichthyological Bulletin	Irregular	January 1956	Rhodes University

Organizations

Organizations	Organizations
• American Elasmobranch Society • American Fisheries Society • American Society of Ichthyologists and Herpetologists • Association of Systematics Collections • Ichthyological Society of Hong Kong (Chinese: 香港魚類學會) • Native Fish Conservancy • Neotropical Ichthyological Association	• North American Native Fishes Association • Panhellenic Society of Technologists Ichthyologists • Society for Integrative and Comparative Biology • Society for Northwestern Vertebrate Biology • Society for the Preservation of Natural History Collections • Southeastern Fishes Council • Southwestern Association of Naturalists • The World Conservation Union

Notable Ichthyologists

Members of this list meet one or more of the following criteria: 1) Author of 50 or more fish taxon names, 2) Author of major reference work in ichthyology, 3) Founder of major journal or museum, 4) Person most notable for other reasons who has also worked in ichthyology.

- Alexander Emanuel Agassiz
- Louis Agassiz
- HIM Emperor Akihito of Japan
- Gerald R. Allen
- Peter Artedi
- Herbert R. Axelrod
- William O. Ayres, California
- Spencer Fullerton Baird

- Tarleton Hoffman Bean
- Lev Berg, Russia
- Henry Bryant Bigelow
- Pieter von Bleeker, East Indies
- Marcus Elieser Bloch
- George Albert Boulenger
- Warren E. Burgess
- Pierre Carbonnier
- Eugenie Clark
- Leonard Compagno
- Edward Drinker Cope
- Georges Cuvier
- Francis Day, India
- Francis Buchanan-Hamilton, Scottish
- Carl H. Eigenmann
- Rosa Smith Eigenmann
- William N. Eschmeyer (pl)
- Barton Warren Evermann
- Henry Weed Fowler
- Joseph Paul Gaimard
- Samuel Garman
- Charles Henry Gilbert
- Theodore Nicholas Gill
- Charles Frédéric Girard
- George Brown Goode
- Albert Günther
- Albert William Herre
- Carl L. Hubbs
- David Starr Jordan
- Maurice Kottelat, Swiss
- Bernard Germain de Lacépède
- Carl Linnaeus
- Seth Eugene Meek
- Ambat Gopalan Kutty Menon (fr) Zoological Survey of India
- George S. Myers
- Joseph S. Nelson *Fishes of the World*

- John Treadwell Nichols, China, founder of *Copeia*
- John Roxborough Norman
- Peter Simon Pallas
- Wilhelm Peters
- Felipe Poey
- Jean René Constant Quoy
- Constantine Samuel Rafinesque
- John E. Randall
- Charles Tate Regan
- John Richardson
- Donn E. Rosen (es)
- Eduard Rüppell
- Johann Gottlob Schneider
- H.M. Smith
- J.L.B. Smith
- Edwin Chapin Starks
- Franz Steindachner
- Shigeho Tanaka
- Ethelwynn Trewavas, English
- Achille Valenciennes
- Johann Julius Walbaum
- Stanley H. Weitzmann
- Francis Willughby

Paleoichthyologists

- Hans C. Bjerring
- Erik Jarvik
- Erik Stensiö

FISH

A **fish** is any member of a group of organisms that consist of all gill-bearing aquatic craniate animals that lack limbs with digits. They form a sister group to the tunicates, together forming the olfactores. Included in this definition are the living hagfish, lampreys, and cartilaginous and bony fish, as well as various extinct related groups. Tetrapods emerged within lobe-finned fishes, so cladistically they are fish as well. However, traditionally fish are rendered obsolete or paraphyletic by excluding the tetrapods (i.e., the amphibians, reptiles, birds and mammals which all descended from within the same ancestry). Because in this manner the term "fish" is defined negatively as a paraphyletic group, it is not considered a formal taxonomic grouping in systematic biology. The traditional term **pisces** (also **ichthyes**) is considered a typological, but not a phylogenetic classification.

The earliest organisms that can be classified as fish were soft-bodied chordates that first appeared during the Cambrian period. Although they lacked a true spine, they possessed notochords which allowed them to be more agile than their invertebrate counterparts. Fish would continue to evolve through the Paleozoic era, diversifying into a wide variety of forms. Many fish of the Paleozoic developed external armor that protected them from predators. The first fish with jaws appeared in the Silurian period, after which many (such as sharks) became formidable marine predators rather than just the prey of arthropods.

Most fish are ectothermic ("cold-blooded"), allowing their body temperatures to vary as ambient temperatures change, though some of the large active swimmers like white shark and tuna can hold a higher core temperature. Fish are abundant in most bodies of water. They can be found in nearly all aquatic environments, from high mountain streams (e.g., char and gudgeon) to the abyssal and even hadal depths of the deepest oceans (e.g., gulpers and anglerfish). With 33,100 described species, fish exhibit greater species diversity than any other group of vertebrates.

Fish are an important resource for humans worldwide, especially as food. Commercial and subsistence fishers hunt fish in wild fisheries (see fishing) or farm them in ponds or in cages in the ocean (see aquaculture). They are also caught by recreational fishers, kept as pets, raised by fishkeepers, and exhibited in public aquaria. Fish have had a role in culture through the ages, serving as deities, religious symbols, and as the subjects of art, books and movies.

Evolution

Dunkleosteus was a gigantic, 10-metre (33 ft) long prehistoric fish of class Placodermi.

Fish do not represent a monophyletic group, and therefore the "evolution of fish" is not studied as a single event.

Early fish from the fossil record are represented by a group of small, jawless, armored fish known as ostracoderms. Jawless fish lineages are mostly extinct. An extant clade, the lampreys may approximate ancient pre-jawed fish. The first jaws are found in Placodermi fossils. The diversity of jawed vertebrates may indicate the evolutionary advantage of a jawed mouth. It is unclear if the advantage of a hinged jaw is greater biting force, improved respiration, or a combination of factors.

Fish may have evolved from a creature similar to a coral-like Sea squirt, whose larvae resemble primitive fish in important ways. The first ancestors of fish may have kept the larval form into adulthood (as some sea squirts do today), although perhaps the reverse is the case.

Taxonomy

Fish are a paraphyletic group: that is, any clade containing all fish also contains the tetrapods, which are not fish. For this reason, groups such as the "Class Pisces" seen in older reference works are no longer used in formal classifications.

Traditional classification divide fish into three extant classes, and with extinct forms sometimes classified within the tree, sometimes as their own classes:

- Class Agnatha (jawless fish).
    - Subclass Cyclostomata (hagfish and lampreys)
    - Subclass Ostracodermi (armoured jawless fish) †
- Class Chondrichthyes (cartilaginous fish)
    - Subclass Elasmobranchii (sharks and rays)
    - Subclass Holocephali (chimaeras and extinct relatives)
- Class Placodermi (armoured fish) †
- Class Acanthodii ("spiny sharks", sometimes classified under bony fishes)†

Leedsichthys (left), of the subclass Actinopterygii, is the largest known fish, with estimates in 2005 putting its maximum size at 16 metres (52 ft)

- Class Osteichthyes (bony fish)
    - Subclass Actinopterygii (ray finned fishes)
    - Subclass Sarcopterygii (fleshy finned fishes, ancestors of tetrapods)

The above scheme is the one most commonly encountered in non-specialist and general works. Many of the above groups are paraphyletic, in that they have given rise to successive groups: Agnathans are ancestral to Chondrichthyes, who again have given rise to Acanthodiians, the ancestors of Osteichthyes. With the arrival of phylogenetic nomenclature, the fishes has been split up into a more detailed scheme, with the following major groups:

- Class Myxini (hagfish)
- Class Pteraspidomorphi † (early jawless fish)
- Class Thelodonti †
- Class Anaspida †
- Class Petromyzontida or Hyperoartia
    - Petromyzontidae (lampreys)
- Class Conodonta (conodonts) †
- Class Cephalaspidomorphi † (early jawless fish)
    - (unranked) Galeaspida †
    - (unranked) Pituriaspida †
    - (unranked) Osteostraci †
- Infraphylum Gnathostomata (jawed vertebrates)
    - Class Placodermi † (armoured fish)
    - Class Chondrichthyes (cartilaginous fish)
    - Class Acanthodii † (spiny sharks)
    - Superclass Osteichthyes (bony fish)
        - Class Actinopterygii (ray-finned fish)
            - Subclass Chondrostei
                - Order Acipenseriformes (sturgeons and paddlefishes)
                - Order Polypteriformes (reedfishes and bichirs).
            - Subclass Neopterygii
                - Infraclass Holostei (gars and bowfins)

- Infraclass Teleostei (many orders of common fish)
- Class Sarcopterygii (lobe-finned fish)
    - Subclass Actinistia (coelacanths)
    - Subclass Dipnoi (lungfish)

† – indicates extinct taxon

Some palaeontologists contend that because Conodonta are chordates, they are primitive fish. For a fuller treatment of this taxonomy, see the vertebrate article.

The position of hagfish in the phylum Chordata is not settled. Phylogenetic research in 1998 and 1999 supported the idea that the hagfish and the lampreys form a natural group, the Cyclostomata, that is a sister group of the Gnathostomata.

The various fish groups account for more than half of vertebrate species. There are almost 28,000 known extant species, of which almost 27,000 are bony fish, with 970 sharks, rays, and chimeras and about 108 hagfish and lampreys. A third of these species fall within the nine largest families; from largest to smallest, these families are Cyprinidae, Gobiidae, Cichlidae, Characidae, Loricariidae, Balitoridae, Serranidae, Labridae, and Scorpaenidae. About 64 families are monotypic, containing only one species. The final total of extant species may grow to exceed 32,500.

Diversity

Examples of the major classes of fish

Agnatha

(Pacific hagfish)

Chondrichthyes

(Horn shark)

Actinopterygii (Brown trout)

Sarcopterygii (Coelacanth)

Fish come in many shapes and sizes. This is a sea dragon, a close relative of the seahorse. Their leaf-like appendages enable them to blend in with floating seaweed.

The term "fish" most precisely describes any non-tetrapod craniate (i.e. an animal with a skull and in most cases a backbone) that has gills throughout life and whose limbs, if any, are in the shape of fins. Unlike groupings such as birds or mammals, fish are not a single clade but a paraphyletic collection of taxa, including hagfishes, lampreys, sharks and rays, ray-finned fish, coelacanths, and lungfish. Indeed, lungfish and coelacanths are closer relatives of tetrapods (such as mammals, birds, amphibians, etc.) than of other fish such as ray-finned fish or sharks, so the last common ancestor of all fish is also an ancestor to tetrapods. As paraphyletic groups are no longer recognised in modern systematic biology, the use of the term "fish" as a biological group must be avoided.

Many types of aquatic animals commonly referred to as "fish" are not fish in the sense given above; examples include shellfish, cuttlefish, starfish, crayfish and jellyfish.

In earlier times, even biologists did not make a distinction – sixteenth century natural historians classified also seals, whales, amphibians, crocodiles, even hippopotamuses, as well as a host of aquatic invertebrates, as fish. However, according to the definition above, all mammals, including cetaceans like whales and dolphins, are not fish. In some contexts, especially in aquaculture, the true fish are referred to as **finfish** (or **fin fish**) to distinguish them from these other animals.

A typical fish is ectothermic, has a streamlined body for rapid swimming, extracts oxygen from water using gills or uses an accessory breathing organ to breathe atmospheric oxygen, has two sets of paired fins, usually one or two (rarely three) dorsal fins, an anal fin, and a tail fin, has jaws, has skin that is usually covered with scales, and lays eggs.

Each criterion has exceptions. Tuna, swordfish, and some species of sharks show some warm-blooded adaptations—they can heat their bodies significantly above ambient water temperature. Streamlining and swimming performance varies from fish such as tuna, salmon, and jacks that can cover 10–20 body-lengths per second to species such as eels and rays that swim no more than 0.5 body-lengths per second. Many groups of freshwater fish extract oxygen from the air as well as from the water using a variety of different structures. Lungfish have paired lungs similar to those of tetrapods, gouramis have a structure called the labyrinth organ that performs a similar function, while many catfish, such as *corydoras* extract oxygen via the intestine or stomach. Body shape and the arrangement of the fins is highly variable, covering such seemingly un-fishlike forms as seahorses, pufferfish, anglerfish, and gulpers. Similarly, the surface of the skin may be naked (as in moray eels), or covered with scales of a variety of different types usually defined as placoid (typical of sharks and rays), cosmoid (fossil lungfish and coelacanths), ganoid (various fossil fish but also living gars and bichirs), cycloid, and ctenoid (these last two are found on most bony fish). There are even fish that live mostly on land. Mudskippers feed and interact with one another on mudflats and go underwater to hide in their burrows. The catfish *Phreatobius cisternarum* lives in underground, phreatic habitats, and a relative lives in waterlogged leaf litter.

Fish range in size from the huge 16-metre (52 ft) whale shark to the tiny 8-millimetre (0.3 in) stout infantfish.

Fish species diversity is roughly divided equally between marine (oceanic) and freshwater ecosystems. Coral reefs in the Indo-Pacific constitute the center of diversity for marine fishes, whereas continental freshwater fishes are most diverse in large river basins of tropical rainforests, especially the Amazon, Congo, and Mekong basins. More than 5,600 fish species inhabit Neotropical freshwaters alone, such that Neotropical fishes represent about 10% of all vertebrate species on the Earth. Exceptionally rich sites in the Amazon basin, such as Cantão State Park, can contain more freshwater fish species than occur in all of Europe.

Anatomy

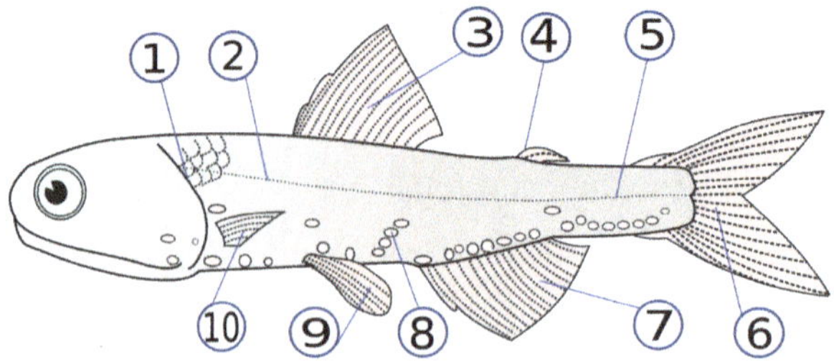

The anatomy of *Lampanyctodes hectoris*

(1) – operculum (gill cover), (2) – lateral line, (3) – dorsal fin, (4) – fat fin, (5) – caudal peduncle, (6) – caudal fin, (7) – anal fin, (8) – photophores, (9) – pelvic fins (paired), (10) – pectoral fins (paired)

Respiration

Gills

Most fish exchange gases using gills on either side of the pharynx. Gills consist of threadlike structures called filaments. Each filament contains a capillary network that provides a large surface area for exchanging oxygen and carbon dioxide. Fish exchange gases by pulling oxygen-rich water through their mouths and pumping it over their gills. In some fish, capillary blood flows in the opposite direction to the water, causing countercurrent exchange. The gills push the oxygen-poor water out through openings in the sides of the pharynx. Some fish, like sharks and lampreys, possess multiple gill openings. However, bony fish have a single gill opening on each side. This opening is hidden beneath a protective bony cover called an operculum.

Juvenile bichirs have external gills, a very primitive feature that they share with larval amphibians.

Air Breathing

Fish from multiple groups can live out of the water for extended periods. Amphibious fish such as the mudskipper can live and move about on land for up to several days, or live in stagnant or otherwise oxygen depleted water. Many such fish can breathe air via a variety of mechanisms. The skin of anguillid eels may absorb oxygen directly. The buccal cavity of the electric eel may breathe air. Catfish of the families Loricariidae, Callichthyidae, and Scoloplacidae absorb air through their digestive tracts. Lungfish, with the exception of the Australian lungfish, and bichirs have paired lungs similar to those of tetrapods and must surface to gulp fresh air through the mouth and pass spent air out through the gills. Gar and bowfin have a vascularized swim bladder that functions in the same way. Loaches, trahiras, and many catfish breathe by passing air through the gut. Mudskippers breathe by absorbing oxygen across the skin (similar to frogs). A number of fish have evolved so-called **accessory breathing organs** that extract oxygen from the air.

Labyrinth fish (such as gouramis and bettas) have a labyrinth organ above the gills that performs this function. A few other fish have structures resembling labyrinth organs in form and function, most notably snakeheads, pikeheads, and the Clariidae catfish family.

Breathing air is primarily of use to fish that inhabit shallow, seasonally variable waters where the water's oxygen concentration may seasonally decline. Fish dependent solely on dissolved oxygen, such as perch and cichlids, quickly suffocate, while air-breathers survive for much longer, in some cases in water that is little more than wet mud. At the most extreme, some air-breathing fish are able to survive in damp burrows for weeks without water, entering a state of aestivation (summertime hibernation) until water returns.

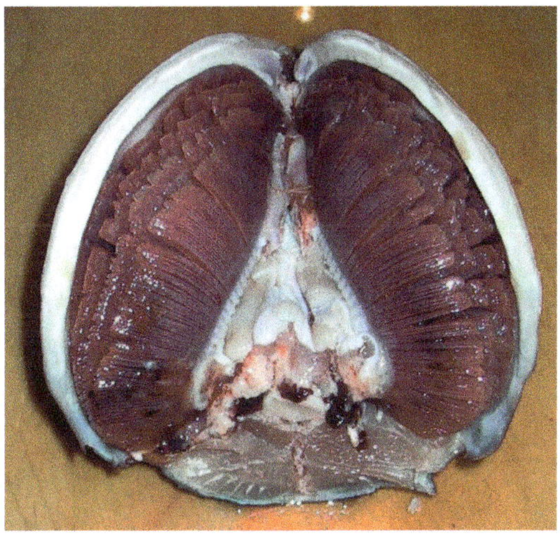

Tuna gills inside the head. The fish head is oriented snout-downwards, with the view looking towards the mouth.

Air breathing fish can be divided into **obligate air breathers** and **facultative air breathers**. Obligate air breathers, such as the African lungfish, *must* breathe air periodically or they suffocate. Facultative air breathers, such as the catfish *Hypostomus plecostomus*, only breathe air if they need to and will otherwise rely on their gills for oxygen. Most air breathing fish are facultative air breathers that avoid the energetic cost of rising to the surface and the fitness cost of exposure to surface predators.

Circulation

Fish have a closed-loop circulatory system. The heart pumps the blood in a single loop throughout the body. In most fish, the heart consists of four parts, including two chambers and an entrance and exit.The first part is the sinus venosus, a thin-walled sac that collects blood from the fish's veins before allowing it to flow to the second part, the atrium, which is a large muscular chamber. The atrium serves as a one-way antechamber, sends blood to the third part, ventricle. The ventricle is another thick-walled, muscular chamber and it pumps the blood, first to the fourth part, bulbus arteriosus, a large tube, and then out of the heart. The bulbus arteriosus connects to the aorta, through which blood flows to the gills for oxygenation.

Digestion

Jaws allow fish to eat a wide variety of food, including plants and other organisms. Fish ingest food through the mouth and break it down in the esophagus. In the stomach, food is further digested and, in many fish, processed in finger-shaped pouches called pyloric caeca, which secrete digestive enzymes and absorb nutrients. Organs such as the liver and pancreas add enzymes and various chemicals as the food moves through the digestive tract. The intestine completes the process of digestion and nutrient absorption.

Excretion

As with many aquatic animals, most fish release their nitrogenous wastes as ammonia. Some of the wastes diffuse through the gills. Blood wastes are filtered by the kidneys.

Saltwater fish tend to lose water because of osmosis. Their kidneys return water to the body. The reverse happens in freshwater fish: they tend to gain water osmotically. Their kidneys produce dilute urine for excretion. Some fish have specially adapted kidneys that vary in function, allowing them to move from freshwater to saltwater.

Scales

The scales of fish originate from the mesoderm (skin); they may be similar in structure to teeth.

Sensory and Nervous System

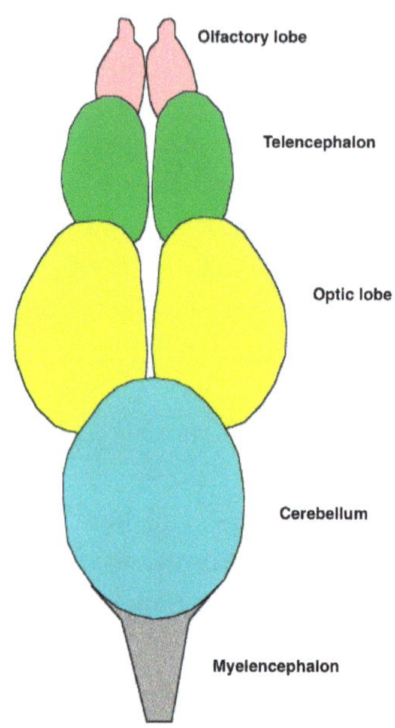

Dorsal view of the brain of the rainbow trout

Central Nervous System

Fish typically have quite small brains relative to body size compared with other vertebrates, typically one-fifteenth the brain mass of a similarly sized bird or mammal. However, some fish have relatively large brains, most notably mormyrids and sharks, which have brains about as massive relative to body weight as birds and marsupials.

Fish brains are divided into several regions. At the front are the olfactory lobes, a pair of structures that receive and process signals from the nostrils via the two olfactory nerves. The olfactory lobes are very large in fish that hunt primarily by smell, such as hagfish, sharks, and catfish. Behind the olfactory lobes is the two-lobed telencephalon, the structural equivalent to the cerebrum in higher vertebrates. In fish the telencephalon is concerned mostly with olfaction. Together these structures form the **forebrain**.

Connecting the forebrain to the **midbrain** is the diencephalon (in the diagram, this structure is below the optic lobes and consequently not visible). The diencephalon performs functions associated with hormones and homeostasis. The pineal body lies just above the diencephalon. This structure detects light, maintains circadian rhythms, and controls color changes.

The midbrain or mesencephalon contains the two optic lobes. These are very large in species that hunt by sight, such as rainbow trout and cichlids.

The **hindbrain** or metencephalon is particularly involved in swimming and balance. The cerebellum is a single-lobed structure that is typically the biggest part of the brain. Hagfish and lampreys have relatively small cerebellae, while the mormyrid cerebellum is massive and apparently involved in their electrical sense.

The **brain stem** or myelencephalon is the brain's posterior. As well as controlling some muscles and body organs, in bony fish at least, the brain stem governs respiration and osmoregulation.

Sense organs

Most fish possess highly developed sense organs. Nearly all daylight fish have color vision that is at least as good as a human's (see vision in fishes). Many fish also have chemoreceptors that are responsible for extraordinary senses of taste and smell. Although they have ears, many fish may not hear very well. Most fish have sensitive receptors that form the lateral line system, which detects gentle currents and vibrations, and senses the motion of nearby fish and prey. Some fish, such as catfish and sharks, have organs that detect weak electric currents on the order of millivolt. Other fish, like the South American electric fishes Gymnotiformes, can produce weak electric currents, which they use in navigation and social communication.

Fish orient themselves using landmarks and may use mental maps based on multiple landmarks or symbols. Fish behavior in mazes reveals that they possess spatial memory and visual discrimination.

Vision

Vision is an important sensory system for most species of fish. Fish eyes are similar to those of terrestrial vertebrates like birds and mammals, but have a more spherical lens. Their retinas generally have both rods and cones (for scotopic and photopic vision), and most species have colour vision. Some fish can see ultraviolet and some can see polarized light. Amongst jawless fish, the lamprey has well-developed eyes, while the hagfish has only primitive eyespots. Fish vision shows adaptation to their visual environment, for example deep sea fishes have eyes suited to the dark environment.

Hearing

Hearing is an important sensory system for most species of fish. Fish sense sound using their lateral lines and their ears.

Capacity for pain

Experiments done by William Tavolga provide evidence that fish have pain and fear responses. For instance, in Tavolga's experiments, toadfish grunted when electrically shocked and over time they came to grunt at the mere sight of an electrode.

In 2003, Scottish scientists at the University of Edinburgh and the Roslin Institute concluded that rainbow trout exhibit behaviors often associated with pain in other animals. Bee venom and acetic acid injected into the lips resulted in fish rocking their bodies and rubbing their lips along the sides and floors of their tanks, which the researchers concluded were attempts to relieve pain, similar to what mammals would do. Neurons fired in a pattern resembling human neuronal patterns.

Professor James D. Rose of the University of Wyoming claimed the study was flawed since it did not provide proof that fish possess "conscious awareness, particularly a kind of awareness that is meaningfully like ours". Rose argues that since fish brains are so different from human brains, fish are probably not conscious in the manner humans are, so that reactions similar to human reactions to pain instead have other causes. Rose had published a study a year earlier arguing that fish cannot feel pain because their brains lack a neocortex. However, animal behaviorist Temple Grandin argues that fish could still have consciousness without a neocortex because "different species can use different brain structures and systems to handle the same functions."

Animal welfare advocates raise concerns about the possible suffering of fish caused by angling. Some countries, such as Germany have banned specific types of fishing, and the British RSPCA now formally prosecutes individuals who are cruel to fish.

Muscular System

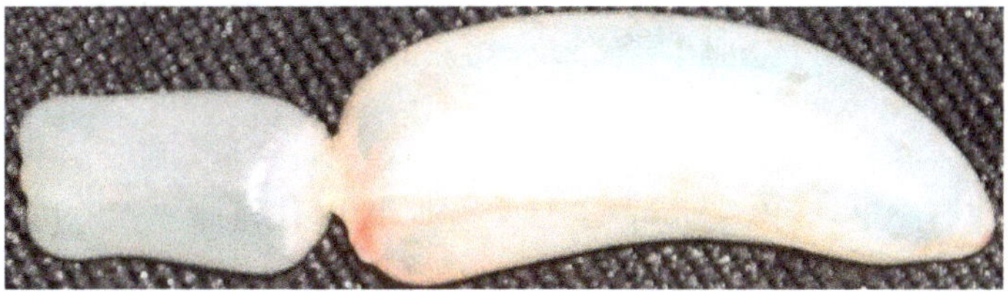

Swim bladder of a rudd (*Scardinius erythrophthalmus*)

Most fish move by alternately contracting paired sets of muscles on either side of the backbone. These contractions form S-shaped curves that move down the body. As each curve reaches the back fin, backward force is applied to the water, and in conjunction with the fins, moves the fish forward. The fish's fins function like an airplane's flaps. Fins also increase the tail's surface area, increasing speed. The streamlined body of the fish decreases the amount of friction from the water. Since body tissue is denser than water, fish must compensate for the difference or they will sink. Many bony fish have an internal organ called a swim bladder that adjusts their buoyancy through manipulation of gases.

A great white shark off Isla Guadalupe

Homeothermy

Although most fish are exclusively ectothermic, there are exceptions.

Certain species of fish maintain elevated body temperatures. Endothermic teleosts (bony fish) are all in the suborder Scombroidei and include the billfishes, tunas, and one species of "primitive" mackerel (*Gasterochisma melampus*). All sharks in the family Lamnidae – shortfin mako, long fin mako, white, porbeagle, and salmon shark – are endothermic, and evidence suggests the trait exists in family Alopiidae (thresher sharks). The degree of endothermy varies from the billfish, which warm only their eyes and brain, to bluefin tuna and porbeagle sharks who maintain body temperatures elevated in excess of 20 °C above ambient water temperatures. *See also gigantothermy*. Endothermy, though metabolically costly, is thought to provide advantages such as increased muscle strength, higher rates of central nervous system processing, and higher rates of digestion.

Reproductive System

Organs: 1. Liver, 2. Gas bladder, 3. Roe, 4. Pyloric caeca, 5. Stomach, 6. Intestine

Fish reproductive organs include testicles and ovaries. In most species, gonads are paired organs of similar size, which can be partially or totally fused. There may also be a range of secondary organs that increase reproductive fitness.

In terms of spermatogonia distribution, the structure of teleosts testes has two types: in the most common, spermatogonia occur all along the seminiferous tubules, while in atherinomorph fish they are confined to the distal portion of these structures. Fish can present cystic or semi-cystic spermatogenesis in relation to the release phase of germ cells in cysts to the seminiferous tubules lumen.

Fish ovaries may be of three types: gymnovarian, secondary gymnovarian or cystovarian. In the first type, the oocytes are released directly into the coelomic cavity and then enter the ostium, then through the oviduct and are eliminated. Secondary gymnovarian ovaries shed ova into the coelom from which they go directly into the oviduct. In the third type, the oocytes are conveyed to the exterior through the oviduct.Gymnovaries are the primitive condition found in lungfish, sturgeon, and bowfin. Cystovaries characterize most teleosts, where the ovary lumen has continuity with the oviduct. Secondary gymnovaries are found in salmonids and a few other teleosts. Oogonia development in teleosts fish varies according to the group, and the determination of oogenesis dynamics allows the understanding of maturation and fertilization processes. Changes in the nucleus, ooplasm, and the surrounding layers characterize the oocyte maturation process.

Postovulatory follicles are structures formed after oocyte release; they do not have endocrine function, present a wide irregular lumen, and are rapidly reabsorbed in a process involving the apoptosis of follicular cells. A degenerative process called follicular atresia reabsorbs vitellogenic oocytes not spawned. This process can also occur, but less frequently, in oocytes in other development stages.

Some fish, like the California sheephead, are hermaphrodites, having both testes and ovaries either at different phases in their life cycle or, as in hamlets, have them simultaneously.

Over 97% of all known fish are oviparous, that is, the eggs develop outside the mother's body. Examples of oviparous fish include salmon, goldfish, cichlids, tuna, and eels. In the majority of these species, fertilisation takes place outside the mother's body, with the male and female fish shedding their gametes into the surrounding water. However, a few oviparous fish practice internal fertilization, with the male using some sort of intromittent organ to deliver sperm into the genital opening of the female, most notably the oviparous sharks, such as the horn shark, and oviparous rays, such as skates. In these cases, the male is equipped with a pair of modified pelvic fins known as claspers.

Marine fish can produce high numbers of eggs which are often released into the open water column. The eggs have an average diameter of 1 millimetre (0.039 in).

Egg of lamprey

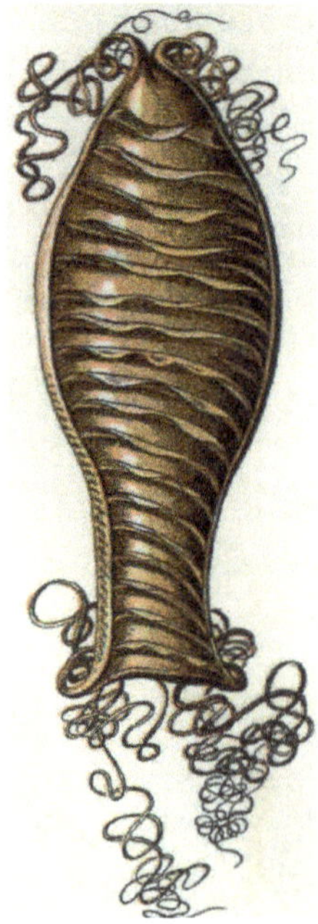

Egg of catshark (mermaids' purse)

Egg of bullhead shark

Egg of chimaera

The newly hatched young of oviparous fish are called larvae. They are usually poorly formed, carry a large yolk sac (for nourishment), and are very different in appearance from juvenile and adult specimens. The larval period in oviparous fish is relatively short (usually only several weeks), and larvae rapidly grow and change appearance and structure (a process termed metamorphosis) to become juveniles. During this transition larvae must switch from their yolk sac to feeding on zooplankton prey, a process which depends on typically inadequate zooplankton density, starving many larvae.

In ovoviviparous fish the eggs develop inside the mother's body after internal fertilization but receive little or no nourishment directly from the mother, depending instead on the yolk. Each embryo develops in its own egg. Familiar examples of ovoviviparous fish include guppies, angel sharks, and coelacanths.

Some species of fish are viviparous. In such species the mother retains the eggs and nourishes the embryos. Typically, viviparous fish have a structure analogous to the placenta seen in mammals connecting the mother's blood supply with that of the embryo. Examples of viviparous fish include the surf-perches, splitfins, and lemon shark. Some viviparous fish exhibit oophagy, in which the developing embryos eat other eggs produced by the mother. This has been observed primarily among sharks, such as the shortfin mako and porbeagle, but is known for a few bony fish as well, such as the halfbeak *Nomorhamphus ebrardtii*. Intrauterine cannibalism is an even more unusual mode of vivipary, in which the largest embryos eat weaker and smaller siblings. This behavior is also most commonly found among sharks, such as the grey nurse shark, but has also been reported for *Nomorhamphus ebrardtii*.

Aquarists commonly refer to ovoviviparous and viviparous fish as livebearers.

Diseases

Like other animals, fish suffer from diseases and parasites. To prevent disease they have a variety of defenses. *Non-specific* defenses include the skin and scales, as well as the mucus layer secreted by the epidermis that traps and inhibits the growth of microorganisms. If pathogens breach these defenses, fish can develop an inflammatory response that increases blood flow to the infected region and delivers white blood cells that attempt to destroy pathogens. Specific defenses respond to particular pathogens recognised by the fish's body, i.e., an immune response. In recent years, vaccines have become widely used in aquaculture and also with ornamental fish, for example furunculosis vaccines in farmed salmon and koi herpes virus in koi.

Some species use cleaner fish to remove external parasites. The best known of these are the Bluestreak cleaner wrasses of the genus *Labroides* found on coral reefs in the Indian and Pacific oceans. These small fish maintain so-called "cleaning stations" where other fish congregate and perform specific movements to attract the attention of the cleaners. Cleaning behaviors have been observed in a number of fish groups, including an interesting case between two cichlids of the same genus, *Etroplus maculatus*, the cleaner, and the much larger *Etroplus suratensis*.

Immune System

Immune organs vary by type of fish. In the jawless fish (lampreys and hagfish), true lymphoid organs are absent. These fish rely on regions of lymphoid tissue within other organs to produce immune cells. For example, erythrocytes, macrophages and plasma cells are produced in the anterior kidney (or pronephros) and some areas of the gut (where granulocytes mature.) They resemble primitive bone marrow in hagfish. Cartilaginous fish (sharks and rays) have a more advanced immune system. They have three specialized organs that are unique to Chondrichthyes; the epigonal organs (lymphoid tissue similar to mammalian bone) that surround the gonads, the Leydig's organ within the walls of their esophagus, and a spiral valve in their intestine. These organs house typical immune cells (granulocytes, lymphocytes and plasma cells). They also possess an identifiable thymus and a well-developed spleen (their most important immune organ) where various lymphocytes, plasma cells and macrophages develop and are stored. Chondrostean fish (sturgeons, paddlefish, and bichirs) possess a major site for the production of granulocytes within a mass that is associated with the meninges (membranes surrounding the central nervous system.) Their heart is frequently covered with tissue that contains lymphocytes, reticular cells and a small number of macrophages. The chondrostean kidney is an important hemopoietic organ; where erythrocytes, granulocytes, lymphocytes and macrophages develop.

Like chondrostean fish, the major immune tissues of bony fish (or teleostei) include the kidney (especially the anterior kidney), which houses many different immune cells. In addition, teleost fish possess a thymus, spleen and scattered immune areas within mucosal tissues (e.g. in the skin, gills, gut and gonads). Much like the mammalian immune system, teleost erythrocytes, neutrophils and granulocytes are believed to reside in the spleen whereas lymphocytes are the major cell type found in the thymus. In 2006, a lymphatic system similar to that in mammals was described in one species of teleost fish, the zebrafish. Although not confirmed as yet, this system presumably will be where naive (unstimulated) T cells accumulate while waiting to encounter an antigen.

B and T lymphocytes bearing immunoglobulins and T cell receptors, respectively, are found in all jawed fishes. Indeed, the adaptive immune system as a whole evolved in an ancestor of all jawed vertebrate.

Conservation

The 2006 IUCN Red List names 1,173 fish species that are threatened with extinction. Included are species such as Atlantic cod, Devil's Hole pupfish, coelacanths, and great white sharks. Because fish live underwater they are more difficult to study than terrestrial animals and plants, and information about fish populations is often lacking. However, freshwater fish seem particularly threatened because they often live in relatively small water bodies. For example, the Devil's Hole pupfish occupies only a single 3 by 6 metres (10 by 20 ft) pool.

Overfishing

A Whale shark, the world's largest fish, is classified as Vulnerable.

Overfishing is a major threat to edible fish such as cod and tuna. Overfishing eventually causes population (known as stock) collapse because the survivors cannot produce enough young to replace those removed. Such **commercial extinction** does not mean that the species is extinct, merely that it can no longer sustain a fishery.

One well-studied example of fishery collapse is the Pacific sardine *Sadinops sagax caerulues* fishery off the California coast. From a 1937 peak of 790,000 long tons (800,000 t) the catch steadily declined to only 24,000 long tons (24,000 t) in 1968, after which the fishery was no longer economically viable.

The main tension between fisheries science and the fishing industry is that the two groups have different views on the resiliency of fisheries to intensive fishing. In places such as Scotland, Newfoundland, and Alaska the fishing industry is a major employer, so governments are predisposed to support it. On the other hand, scientists and conservationists push for stringent protection, warning that many stocks could be wiped out within fifty years.

Habitat Destruction

A key stress on both freshwater and marine ecosystems is habitat degradation including water pollution, the building of dams, removal of water for use by humans, and the introduction of exotic species. An example of a fish that has become endangered because of habitat change is the pallid sturgeon, a North American freshwater fish that lives in rivers damaged by human activity.

Exotic Species

Introduction of non-native species has occurred in many habitats. One of the best studied examples is the introduction of Nile perch into Lake Victoria in the 1960s. Nile perch gradually exterminated the lake's 500 endemic cichlid species. Some of them survive now in captive breeding programmes, but others are probably extinct. Carp, snakeheads, tilapia, European perch, brown trout, rainbow trout, and sea lampreys are other examples of fish that have caused problems by being introduced into alien environments.

IMPORTANCE TO HUMANS

Economic Importance

These fish-farming ponds were created as a cooperative project in a rural village.\

Throughout history, humans have utilized fish as a food source. Historically and today, most fish protein has come by means of catching wild fish. However, aquaculture, or fish farming, which has been practiced since about 3,500 BCE. in China, is becoming increasingly important in many nations. Overall, about one-sixth of the world's protein is estimated to be provided by fish. That proportion is considerably elevated in some developing nations and regions heavily dependent on the sea. In a similar manner, fish have been tied to trade.

Catching fish for the purpose of food or sport is known as fishing, while the organized effort by humans to catch fish is called a fishery. Fisheries are a huge global business and provide income for millions of people. The annual yield from all fisheries worldwide is about 154 million tons, with popular species including herring, cod, anchovy, tuna, flounder, and salmon. However, the term fishery is broadly applied, and includes more organisms than just fish, such as mollusks and crustaceans, which are often called "fish" when used as food.

Recreation

Fish have been recognized as a source of beauty for almost as long as used for food, appearing in cave art, being raised as ornamental fish in ponds, and displayed in aquariums in homes, offices, or public settings.

Recreational fishing is fishing for pleasure or competition; it can be contrasted with commercial fishing, which is fishing for profit. The most common form of recreational fishing is done with a rod, reel, line, hooks and any one of a wide range of baits. Angling is a method of fishing, specifically the practice of catching fish by means of an "angle" (hook). Anglers must select the right hook, cast accurately, and retrieve at the right speed while considering water and weather conditions, species, fish response, time of the day, and other factors.

Culture

Fish feature prominently in art and literature, in movies such as *Finding Nemo* and books such as *The Old Man and the Sea*. Large fish, particularly sharks, have frequently been the subject of horror movies and thrillers, most notably the novel *Jaws*, which spawned a series of films of the same name that in turn inspired similar films or parodies such as *Shark Tale* and *Snakehead Terror*. Piranhas are shown in a similar light to sharks in films such as *Piranha*; however, contrary to popular belief, the red-bellied piranha is actually a generally timid scavenger species that is unlikely to harm humans. In the Book of Jonah a "great fish" swallowed Jonah the Prophet. Legends of half-human, half-fish mermaids have featured in folklore, like the stories of Hans Christian Andersen.

Avatar of Vishnu as a Matsya

The ichthus is a Christian symbol of a fish signifying that the person who uses it is a Christian.

Fish themes have symbolic significance in many religions. The fish is used often as a symbol by Christians to represent Jesus, or Christianity in general; the gospels also refer to "fishers of men" and feeding the multitude. In the dhamma of Buddhism the fish symbolize happiness as they have complete freedom of movement in the water. Often drawn in the form of carp which are regarded in the Orient as sacred on account of their elegant beauty, size and life-span. Among the deities said to take the form of a fish are Ika-Roa of the Polynesians, Dagon of various ancient Semitic peoples, the shark-gods of Hawaii and Matsya of the Hindus.

The astrological symbol Pisces is based on a constellation of the same name, but there is also a second fish constellation in the night sky, Piscis Austrinus.

TERMINOLOGY

Shoal or School

These goldband fusiliers are schooling because their swimming is synchronized

A random assemblage of fish merely using some localised resource such as food or nesting sites is known simply as an *aggregation*. When fish come together in an interactive, social grouping, then they may be forming either a *shoal* or a *school* depending on the degree of organisation. A *shoal* is a loosely organised group where each fish swims and forages independently but is attracted to other members of the group and adjusts its behaviour, such as swimming speed, so that it remains close to the other members of the group. *Schools* of fish are much more tightly organised, synchronising their swimming so that all fish move at the same speed and in the same direction. Shoaling and schooling behaviour is believed to provide a variety of advantages.

Examples:

- Cichlids congregating at lekking sites form an *aggregation*.
- Many minnows and characins form *shoals*.
- Anchovies, herrings and silversides are classic examples of *schooling* fish.

While the words "school" and "shoal" have different meanings within biology, they are often treated as synonyms by non-specialists, with speakers of British English using "shoal" to describe any grouping of fish, and speakers of American English often using "school" just as loosely.

Fish or Fishes

Though often used interchangeably, in biology these words have different meanings. *Fish* is used as a singular noun, or as a plural to describe multiple individuals from a single species. *Fishes* is used to describe different species or species groups. Thus a pond that contained a single species might be said to contain 120 fish. But if the pond contained a total of 120 fish from three different species, it would be said to contain three fishes. The distinction is similar to that between people and peoples.

Finfish

- In biology, the term *fish* is most strictly used to describe any animal with a backbone that has gills throughout life and has limbs, if any, in the shape of fins. Many types of aquatic animals with common names ending in "fish" are not fish in this sense; examples include shellfish, cuttlefish, starfish, crayfish and jellyfish. In earlier times, even biologists did not make a distinction—sixteenth century natural historians classified also seals, whales, amphibians, crocodiles, even hippopotamuses, as well as a host of aquatic invertebrates, as fish.
- In fisheries, the term *fish* is used as a collective term, and includes mollusks, crustaceans and any aquatic animal which is harvested.
- The strict biological definition of a fish, above, is sometimes called a *true fish*. True fish are also referred to as *finfish* or *fin fish* to distinguish them from other aquatic life harvested in fisheries or aquaculture.

REFERENCES

1. *oldman, K.J. (1997). "Regulation of body temperature in the white shark, Carcharodon carcharias". Journal of Comparative Physiology. B Biochemical Systemic and Environmental Physiology.* ***167*** *(6): 423–429. doi:10.1007/s003600050092. Retrieved 12 October 2011.*
2. *Carey, F.G.; Lawson, K.D. (February 1973). "Temperature regulation in free-swimming bluefin tuna". Comparative Biochemistry and Physiology A.* ***44*** *(2): 375–392. doi:10.1016/0300-9629(73)90490-8.*
3. *"FishBase". FishBase. April 2015. Retrieved 29 August 2015.*
4. *"Monster fish crushed opposition with strongest bite ever". Smh.com.au. 30 November 2006. Retrieved 26 February 2013.*
5. G. Lecointre & H. Le Guyader, 2007, *The Tree of Life: A Phylogenetic Classification*, Harvard University Press Reference Library
6. Romer, A.S. & T.S. Parsons. 1977. *The Vertebrate Body.* 5th ed. Saunders, Philadelphia. (6th ed. 1985)
7. Benton, M. J. (1998) The quality of the fossil record of vertebrates. Pp. 269–303, in Donovan, S. K. and Paul, C. R. C. (eds), The adequacy of the fossil record, Fig. 2. Wiley, New York, 312 pp.
8. Shigehiro Kuraku, Daisuke Hoshiyama, Kazutaka Katoh, Hiroshi Suga, Takashi

Miyata (1999) *Monophyly of Lampreys and Hagfishes Supported by Nuclear DNA–Coded Genes* J Mol Evol (1999) 49:729–735

9. J. Mallatt, J. Sullivan (1998) *28S and 18S rDNA sequences support the monophyly of lampreys and hagfishes* Molecular Biology and Evolution V 15, Issue 12, pp 1706–1718
10. Nelson 2006, pp. 4–5
11. Nelson 2006, p. 3
12. Nelson 2006, p. 2
13. Helfman, Collette & Facey 1997, p. 3
14. Tree of life web project – Chordates.
15. *Cleveland P. Hickman, Jr.; Larry S. Roberts; Allan L. Larson (2001). Integrated Principles of Zoology. McGraw-Hill Publishing Co. ISBN 0-07-290961-7.*
16. Helfman, Collette & Facey 1997, p. 103
17. Helfman, Collette & Facey 1997, pp. 53 57
18. Helfman, Collette & Facey 1997, pp. 33–36
19. Froese, Rainer and Pauly, Daniel, eds. (2006). "*Periophthalmus barbarus*" in FishBase. November 2006 version.
20. Froese, Rainer and Pauly, Daniel, eds. (2006). "*Phreatobius cisternarum*" in FishBase. November 2006 version.
21. *Planet Catfish. "Cat-eLog: Heptapteridae: Phreatobius: Phreatobiussp. (1)". Planet Catfish. Retrieved 26 November 2006.*
22. Estudo das Espécies Ícticas do Parque Estadual do Cantão, fish species survey of Cantão (in Portuguese)
23. *"Modifications of the Digestive Tract for Holding Air in Loricariid and Scoloplacid Catfishes" (PDF). Copeia (3): 663–675. 1998. doi:10.2307/1447796. Retrieved 25 June 2009.*
24. *Setaro, John F. (1999). Circulatory System. Microsoft Encarta 99.*
25. Helfman, Collette & Facey 1997, pp. 48–49
26. Helfman, Collette & Facey 1997, p. 191
27. *Orr, James (1999). Fish. Microsoft Encarta 99. ISBN 0-8114-2346-8.*
28. Albert, J.S., and W.G.R. Crampton. 2005. Electroreception and electrogenesis. pp. 431–472 in The Physiology of Fishes, 3rd Edition. D.H. Evans and J.B. Claiborne (eds.). CRC Press.
29. *Sciences, Journal of Undergraduate Life. "Appropriate maze methodology to study learning in fish" (PDF). Retrieved 28 May 2009.*
30. *Campbell, Neil A.; Reece, Jane B. (2005). Biology (Seventh ed.). San Francisco, California: Benjamin Cummings.*
31. Dunayer, Joan, "Fish: Sensitivity Beyond the Captor's Grasp," The Animals' Agenda, July/August 1991, pp. 12–18

32. *Kirby, Alex (30 April 2003). "Fish do feel pain, scientists say". BBC News. Retrieved 4 January 2010.*
33. *Grandin, Temple; Johnson, Catherine (2005). Animals in Translation. New York, New York: Scribner. pp. 183–184. ISBN 0-7432-4769-8.*
34. *"Rose, J.D. 2003. A Critique of the paper: "Do fish have nociceptors: Evidence for the evolution of a vertebrate sensory system"" (PDF). Retrieved 21 May 2011.*
35. *Rose, James D. (2002). "Do Fish Feel Pain?". Retrieved 27 September 2007.*
36. Leake, J. "Anglers to Face RSPCA Check," The Sunday Times – Britain, 14 March 2004
37. *Block, BA; Finnerty, JR (1993). "Endothermy in fishes: a phylogenetic analysis of constraints, predispositions, and selection pressures"(PDF). Environmental Biology of Fishes.* ***40*** *(3): 283–302. doi:10.1007/BF00002518.*
38. *Guimaraes-Cruz, Rodrigo J., Rodrigo J.; Santos, José E. dos; Santos, Gilmar B. (July–September 2005). "Gonadal structure and gametogenesis of Loricaria lentiginosa Isbrücker (Pisces, Teleostei, Siluriformes)". Rev. Bras. Zool.* ***22*** *(3): 556–564. doi:10.1590/S0101-81752005000300005. ISSN 0101-8175.*
39. *Brito, M.F.G.; Bazzoli, N. (2003). "Reproduction of the surubim catfish (Pisces, Pimelodidae) in the São Francisco River, Pirapora Region, Minas Gerais, Brazil". Arquivo Brasileiro de Medicina Veterinária e Zootecnia.* ***55*** *(5): 624–633. doi:10.1590/S0102-09352003000500018. ISSN 0102-0935.*
40. Peter Scott: *Livebearing Fishes*, p. 13. Tetra Press 1997. ISBN 1-56465-193-2
41. Meisner, A & Burns, J: Viviparity in the Halfbeak Genera *Dermogenys* and *Nomorhamphus* (Teleostei: Hemiramphidae)" *Journal of Morphology* 234, pp. 295–317, 1997
42. Helfman, Collette & Facey 1997, pp. 95–96
43. R. C. Cipriano (2001), Furunculosis And Other Diseases Caused By *Aeromonas salmonicida*. Fish Disease Leaflet 66. U.S. Department of the Interior.
44. *Hartman, K H; et al. (2004). "Koi Herpes Virus (KHV) Disease: Fact Sheet VM-149" (PDF). University of Florida Institute of Food and Agricultural Sciences.*
45. Helfman, Collette & Facey 1997, p. 380
46. *Wyman, Richard L.; Ward, Jack A. (1972). "A Cleaning Symbiosis between the Cichlid Fishes Etroplus maculatus and Etroplus suratensis. I. Description and Possible Evolution". Copeia.* ***1972*** *(4): 834–838. doi:10.2307/1442742.*
47. A.G. Zapata, A. Chiba and A. Vara. *Cells and tissues of the immune system of fish.* In: The Fish Immune System: Organism, Pathogen and Environment. Fish Immunology Series. (eds. G. Iwama and T.Nakanishi,), New York, Academic Press, 1996, pp. 1–55.
48. D.P. Anderson. *Fish Immunology*. (S.F. Snieszko and H.R. Axelrod, eds), Hong Kong: TFH Publications, Inc. Ltd., 1977.
49. *Chilmonczyk, S. (1992). "The thymus in fish: development and possible function in the immune response".* ***2****. Annual Review of Fish Diseases: 181–200.*

50. *Hansen, J.D.; Zapata, A.G. (1998). "Lymphocyte development in fish and amphibians". **166**. Immunological Reviews: 199–220.*
51. *Kucher; et al. (2006). "Development of the zebrafish lymphatic system requires VegFc signalling". **16**. Current Biology: 1244–1248.*
52. Flajnik, M. F., and M. Kasahara. "Origin and evolution of the adaptive immune system: genetic events and selective pressures." Nature Reviews Genetics 11.1, 47-59 (2009).
53. *"Table 1: Numbers of threatened species by major groups of organisms (1996–2004)". iucnredlist.org. Archived from the originalon 30 June 2006. Retrieved 18 January 2006.*
54. *"Gadus morhua (Atlantic Cod)". Iucnredlist.org. Retrieved 21 May 2011.*
55. *"Cyprinodon diabolis (Devils Hole Pupfish)". Iucnredlist.org. Retrieved 21 May 2011.*
56. *"Latimeria chalumnae (Coelacanth, Gombessa)". Iucnredlist.org. Retrieved 21 May 2011.*
57. *"Carcharodon carcharias (Great White Shark)". Iucnredlist.org. Retrieved 21 May 2011.*
58. Helfman, Collette & Facey 1997, pp. 449–450
59. *"Call to halt cod ‹over-fishing'". BBC News. 5 January 2007. Retrieved 18 January 2006.*
60. *"Tuna groups tackle overfishing". BBC News. 26 January 2007. Retrieved 18 January 2006.*
61. Helfman, Collette & Facey 1997, p. 462
62. *"UK ‹must shield fishing industry'". BBC News. 3 November 2006. Retrieved 18 January 2006.*
63. *"EU fish quota deal hammered out". BBC News. 21 December 2006. Retrieved 18 January 2006.*
64. *"Ocean study predicts the collapse of all seafood fisheries by 2050". Retrieved 13 January 2006.*
65. *"Atlantic bluefin tuna could soon be commercially extinct". Archived from the original on 30 April 2007. Retrieved 18 January 2006.*
66. Helfman, Collette & Facey 1997, p. 463
67. *"Threatened and Endangered Species: Pallid SturgeonScaphirhynchus Fact Sheet". Archived from the original on 26 November 2005. Retrieved 18 March 2016.*
68. *Spinney, Laura (4 August 2005). "The little fish fight back". The Guardian. London. Retrieved 18 January 2006.*
69. *"Stop That Fish!". The Washington Post. 3 July 2002. Retrieved 26 August 2007.*
70. *Spalding, Mark (July 11, 2013). "Sustainable Ancient Aquaculture". National Geographic. Retrieved 13 August 2015.*

71. *Helfman, Gene S. (2007). Fish Conservation: A Guide to Understanding and Restoring Global Aquatic Biodiversity and Fishery Resources. Island Press. p. 11. ISBN 1597267600.*

72. *"World Review of Fisheries and Aquaculture" (PDF). fao.org. Food and Agriculture Organization of the United Nations. Retrieved 13 August 2015.*

73. *Zollinger, Sue Anne (3 July 2009). "Piranha–Ferocious Fighter or Scavenging Softie?". A Moment of Science. Indiana Public Media. Retrieved 1 November 2015.*

74. *Coffman, Elesha (August 8, 2008). "What is the origin of the Christian fish symbol?". Christianity Today. Retrieved 13 August 2015.*

75. *"Piscis Austrinus". allthesky.com. The Deep Photographic Guide to the Constellations. Retrieved 1 November 2015.*

76. *Helfman, G.; Collette, B.; Facey, D. (1997). The Diversity of Fishes. Blackwell Publishing. p. 375. ISBN 0-86542-256-7.*

77. *Pauly, Daniel (13 May 2004). "Fish(es)". Darwin's Fishes: An Encyclopedia of Ichthyology, Ecology, and Evolution. Cambridge University Press. p. 77. ISBN 978-1-139-45181-9.*

78. Nelson, Joseph S and Paetz, Martin Joseph (1992) *The Fishes of Alberta* page 400, University of Alberta. ISBN 9780888642363.

79. Helfman, Collette & Facey 1997, p. 5

80. *Nelson, Joseph S. (2006). Fishes of the World. John Wiley & Sons, Inc. p. 2. ISBN 0-471-25031-7.*

81.

CHAPTER 5

Mammalogy

Chapter Contents

- **Introduction**
- **Mammal**

Mammalogy, scientific study of mammals. Interest in nonhuman mammals dates far back in prehistory, and the modern science of mammalogy has its broad foundation in the knowledge of mammals possessed by primitive peoples.

INTRODUCTION

In zoology, mammalogy is the study of mammals – a class of vertebrates with characteristics such as homeothermic metabolism, fur, four-chambered hearts, and complex nervous systems. Mammalogy has also been known as "mastology," "theriology," and "therology." There are about 4,200 different species of animals which are considered mammals. The major branches to study in the mammalogy career include natural history, taxonomy and systematics, anatomy and physiology, ethology, ecology, and management and control. The approximate salary of a mammalogist varies from $20,000 to $60,000 a year, depending on their experience. Mammalogists are typically involved in activities such as conducting research, managing personnel, and writing proposals.

Siberian tiger

Mammalogy branches off into other taxonomically-oriented disciplines such as primatology (study of primates), and cetology (study of cetaceans).

List of Mammalogy Journals

This is a list of scientific journals broadly serving mammalogists. In addition, many other more general zoology, ecology and evolution, or conservation journals also deal with mammals, and several journals are specific to only certain taxonomic groups of mammals.

Journal	Organization	Publication Schedule	Impact Factor	First Issue	Open Access	Page Charges	Online ISSN	Print ISSN	Former Name
Journal of Mammalogy	American Society of Mammalogists	Bimonthly	2.308	1919	Option	Yes	1545-1542	0022-2372	none
Mammal Review	The Mammal Society	Quarterly	3.919	1970	Option	No	1365-2907	-	none
Mammalian Biology	German Society for Mammalian Biology	Bimonthly	1.337	1935	Option	No	1616-5047	-	Zeitschrift für Säugetierkunde
Mammalia	-	Quarterly	0.824	1936	Option	No	1864-1547	-	none
Mammal Research	Polish Academy of Sciences	Quarterly	1.161	1954	Option	No	2199-241X	2199-2401	Acta Theriologica
Mammal Study	Mammal Society of Japan	Quarterly	0.426	1959	No	No	1348-6160	1343-4152	Journal of the Mammalogical Society of Japan
Hystrix, Italian Journal of Mammalogy	Italian Theriological Association	Quarterly	0.593	1986	Yes	No	1825-5272	-	none
Galemys, Spanish Journal of Mammalogy	Spanish Society of Mammalogists	Annually	-	1988	Yes	No	2254-8408	-	none
Lutra	Dutch Mammal Society	Biannually	-	1957	Yes	No	-	-	none
Australian Mammalogy	Australian Mammal Society	Biannually	-	1972	No	No	1836-7402	0310-0049	none
Acta Theriologica Sinica	Mammalogical Society of China	Quarterly	-	1981	Yes	No	1000-1050	-	none

MAMMAL

Mammals are any vertebrates within the class **Mammalia**, a clade of endothermic amniotes distinguished from reptiles and birds by the possession of a neocortex (a region of the brain), hair, three middle ear bones and mammary glands. The sister group of Mammals may be the extinct *Haldanodon.* The Mammals represent the only living Synapsida, which together with the Sauropsida form the Amniota clade. The Mammals consist of the Australosphenida including monotrema and the theria. Mammals include the largest animals on the planet, the great whales, as well as some of the most intelligent, such as elephants, primates and cetaceans. The basic body type is a terrestrial quadruped, but some mammals are adapted for life at sea, in the air, in trees, underground or on two legs. The largest group of mammals, the placentals, have a placenta, which enables the feeding of the fetus during gestation. Mammals range in size from the 30–40 mm (1.2–1.6 in) bumblebee bat to the 30-meter (98 ft) blue whale. With the exception of the five species of monotreme (egg-laying mammals), all modern mammals give birth to live young. Most mammals, including the six most species-rich orders, belong to the placental group. The three largest orders in number of species are Rodentia: mice, rats, porcupines, beavers, capybaras and other gnawing mammals; Chiroptera: bats; and Soricomorpha: shrews, moles and solenodons. The next three biggest orders, depending on the biological classification scheme used, are the Primates including the great apes and monkeys; the Cetartiodactyla including whales and even-toed ungulates; and the Carnivora which includes cats, dogs, weasels, bears and seals.

The word "mammal" is modern, from the scientific name *Mammalia*, coined by Carl Linnaeus in 1758, derived from the Latin *mamma* ("teat, pap"). All female mammals nurse their young with milk, secreted from the mammary glands. According to *Mammal Species of the World*, 5,416 species were known in 2006. These were grouped in 1,229 genera, 153 families and 29 orders. In 2008 the International Union for Conservation of Nature (IUCN) completed a five-year, 1,700-scientist Global Mammal Assessment for its IUCN Red List, which counted 5,488 species. In some classifications, extant mammals are divided into two subclasses: the Prototheria, that is, the order Monotremata; and the Theria, or the infraclasses Metatheria and Eutheria. The marsupials constitute the crown group of the Metatheria, and include all living metatherians as well as many extinct ones; the placentals are the crown group of the Eutheria. While mammal classification at the family level has been relatively stable, several contending classifications regarding the higher levels—subclass, infraclass and order, especially of the marsupials—appear in contemporaneous literature. Much of the changes reflect the advances of cladistic analysis and molecular genetics. Findings from molecular genetics, for example, have prompted adopting new groups, such as the Afrotheria, and abandoning traditional groups, such as the Insectivora. The early synapsid mammalian ancestors were sphenacodont pelycosaurs, a group that produced the non-mammalian *Dimetrodon*. At the end of the Carboniferous period, this group diverged from the sauropsid line that led to today's reptiles and birds. The line following the stem group Sphenacodontia split-off several diverse groups of non-mammalian synapsids—sometimes referred to as mammal-like reptiles—before giving rise to the proto-mammals (Therapsida) in the early Mesozoic era. The modern mammalian orders arose in the Paleogene and Neogene periods of the Cenozoic era, after the extinction of non-avian dinosaurs, and have been among the dominant terrestrial animal groups from

66 million years ago to the present. In human culture, domesticated mammals played a major role in the Neolithic revolution, causing farming to replace hunting and gathering, and leading to a major restructuring of human societies with the first civilizations. They provided, and continue to provide, power for transport and agriculture, as well as various commodities such as meat, dairy products, wool, and leather. Mammals are hunted or raced for sport, and are used as model organisms in science. Mammals have been depicted in art since Palaeolithic times, and appear in literature, film, mythology, and religion.

Classification

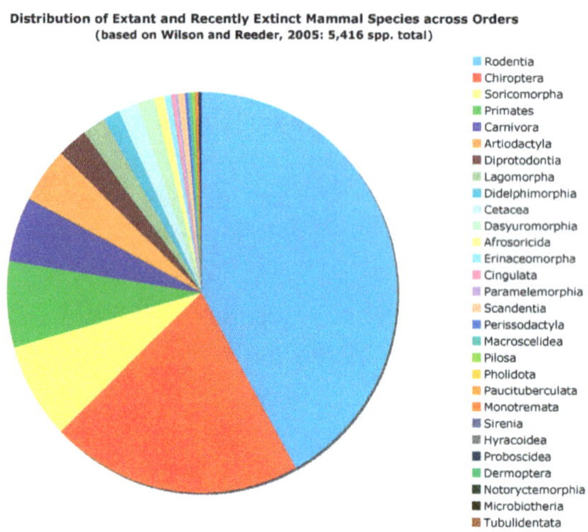

The orders Rodentia (blue), Chiroptera (red) and Soricomorpha (yellow) together comprise over 70% of mammal species.

Mammal classification has been through several iterations since Carl Linnaeus initially defined the class. No classification system is universally accepted; McKenna & Bell (1997) and Wilson & Reader (2005) provide useful recent compendiums. George Gaylord Simpson's "Principles of Classification and a Classification of Mammals" (AMNH *Bulletin* v. 85, 1945) provides systematics of mammal origins and relationships that were universally taught until the end of the 20th century. Since Simpson's classification, the paleontological record has been recalibrated, and the intervening years have seen much debate and progress concerning the theoretical underpinnings of systematization itself, partly through the new concept of cladistics. Though field work gradually made Simpson's classification outdated, it remains the closest thing to an official classification of mammals.

Definitions

The word "mammal" is modern, from the scientific name *Mammalia* coined by Carl Linnaeus in 1758, derived from the Latin *mamma* ("teat, pap"). In an influential 1988 paper, Timothy Rowe defined Mammalia phylogenetically as the crown group of mammals, the clade consisting of the most recent common ancestor of living monotremes (echidnas and platypuses) and therian mammals (marsupials and placentals) and all descendants of that ancestor. Since this ancestor lived in the Jurassic period, Rowe's definition excludes all animals from the earlier Triassic, despite the fact that Triassic fossils in the Haramiyida have been referred to the Mammalia since the mid-19th century. If Mammalia is considered as the crown group, its origin can be roughly dated as the first known appearance of animals more closely related to some extant mammals than to others. *Ambondro* is more closely related to monotremes than to therian mammals while *Amphilestes* and *Amphitherium* are more closely related to the therians; as fossils of all three genera are dated about 167 million years ago in the Middle Jurassic, this is a reasonable estimate for the appearance of the crown group.

T. S. Kemp has provided a more traditional definition: "synapsids that possess a dentary–squamosal jaw articulation and occlusion between upper and lower molars with a transverse component to the movement" or, equivalently in Kemp's view, the clade originating with the last common ancestor of *Sinoconodon* and living mammals. The earliest known synapsid satisfying Kemp's definitions is *Tikitherium*, dated 225 Ma, so the appearance of mammals in this broader sense can be given this Late Triassic date.

McKenna/Bell classification

In 1997, the mammals were comprehensively revised by Malcolm C. McKenna and Susan K. Bell, which has resulted in the McKenna/Bell classification. Their 1997 book, *Classification of Mammals above the Species Level*, is a comprehensive work on the systematics, relationships and occurrences of all mammal taxa, living and extinct, down through the rank of genus, though molecular genetic data challenge several of the higher level groupings. The authors worked together as paleontologists at the American Museum of Natural History, New York. McKenna inherited the project from Simpson and, with Bell, constructed a completely updated hierarchical system, covering living and extinct taxa that reflects the historical genealogy of Mammalia.

Extinct groups are represented by a dagger (†).

Class Mammalia

- **Subclass Prototheria**: monotremes: echidnas and the platypus
- **Subclass Theriiformes**: live-bearing mammals and their prehistoric relatives
- Infraclass †Allotheria: multituberculates
- Infraclass †Eutriconodonta: eutriconodonts
- Infraclass Holotheria: modern live-bearing mammals and their prehistoric relatives
- Superlegion †Kuehneotheria
- Supercohort Theria: live-bearing mammals
- Cohort Marsupialia: marsupials
- Magnorder Australidelphia: Australian marsupials and the monito del monte
- Magnorder Ameridelphia: New World marsupials. Now considered paraphyletic,

with shrew opossums being closer to australidelphians.

- Cohort Placentalia: placentals
- Magnorder Xenarthra: xenarthrans
- Magnorder Epitheria: epitheres
- Superorder Anagalida: lagomorphs, rodents and elephant shrews
- Superorder Ferae: carnivorans, pangolins, †creodonts and relatives
- Superorder Lipotyphla: insectivorans
- Superorder Archonta: bats, primates, colugos and treeshrews
- Superorder Ungulata: ungulates
- Order Tubulidentata *incertae sedis*: aardvark
- Mirorder Eparctocyona: †condylarths, whales and artiodactyls (even-toed ungulates)
- Mirorder †Meridiungulata: South American ungulates
- Mirorder Altungulata: perissodactyls (odd-toed ungulates), elephants, manatees and hyraxes

Molecular classification of placentals

Molecular studies based on DNA analysis have suggested new relationships among mammal families over the last few years. Most of these findings have been independently validated by retrotransposon presence/absence data.

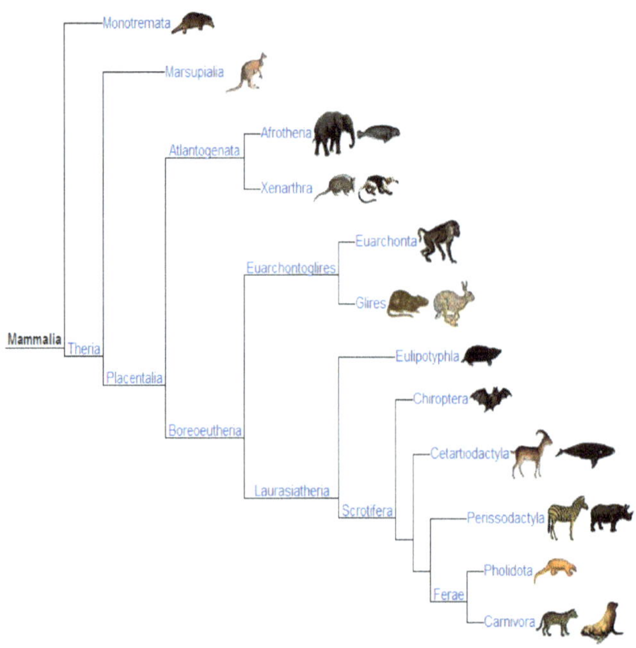

Cladogram based on Tarver *et al.* (2016)

Classification systems based on molecular studies reveal three major groups or lineages of placental mammals—Afrotheria, Xenarthra and Boreoeutheria—which diverged in the Cretaceous. The relationships between these three lineages is contentious, and all three possible different hypotheses have been proposed with respect to which group is basal. These hypotheses are Atlantogenata (basal Boreoeutheria), Epitheria (basal Xenarthra) and Exafroplacentalia (basal Afrotheria).Boreoeutheria in turn contains two major lineages—Euarchontoglires and Laurasiatheria.

Estimates for the divergence times between these three placental groups range from 105 to 120 million years ago, depending on type of DNA (such as nuclear or mitochondrial) and varying interpretations of paleogeographic data.

Group I: Superorder Afrotheria

- Clade Afroinsectiphilia
- Order Macroscelidea: elephant shrews (Africa)
- Order Afrosoricida: tenrecs and golden moles (Africa)
- Order Tubulidentata: aardvark (Africa south of the Sahara)
- Clade Paenungulata
- Order Hyracoidea: hyraxes or dassies (Africa, Arabia)
- Order Proboscidea: elephants (Africa, Southeast Asia)
- Order Sirenia: dugong and manatees (cosmopolitan tropical)

Group II: Superorder Xenarthra

- Order Pilosa: sloths and anteaters (neotropical)
- Order Cingulata: armadillos and extinct relatives (Americas)

Group III: Magnaorder Boreoeutheria

- Superorder: Euarchontoglires (Supraprimates)
- Grandorder Euarchonta
- Order Scandentia: treeshrews (Southeast Asia).
- Order Dermoptera: flying lemurs or colugos (Southeast Asia)
- Order Primates: lemurs, bushbabies, monkeys, apes, humans (cosmopolitan)
- Grandorder Glires
- Order Lagomorpha: pikas, rabbits, hares (Eurasia, Africa, Americas)
- Order Rodentia: rodents (cosmopolitan)
- Superorder: Laurasiatheria
- Order Eulipotyphla: shrews, hedgehogs, moles, solenodons
- *Clade* Ferungulata
- Order Cetartiodactyla: cetaceans (whales, dolphins and porpoises) and even-toed ungulates, including pigs, cattle, deer and giraffes

- *Clade* Pegasoferae
- Order Chiroptera: bats (cosmopolitan)
- *Clade* Zooamata
- Order Perissodactyla: odd-toed ungulates, including horses, donkeys, zebras, tapirs and rhinoceroses
- *Clade* Ferae
- Order Pholidota: pangolins or scaly anteaters (Africa, South Asia)
- Order Carnivora: carnivores (cosmopolitan), including cats and dogs

Statistics

Most mammals, including the six most species-rich orders, belong to the placental group. The three largest orders in numbers of species are Rodentia: mice, rats, porcupines, beavers, capybaras and other gnawing mammals; Chiroptera: bats; and Soricomorpha: shrews, moles and solenodons. The next three biggest orders, depending on the biological classification scheme used, are the Primates including the great apes and monkeys; the Cetartiodactyla including whales and even-toed ungulates; and the Carnivora which includes cats, dogs, weasels, bears and seals. According to *Mammal Species of the World*, 5,416 species were known in 2006. These were grouped in 1,229 genera, 153 families and 29 orders. In 2008 the International Union for Conservation of Nature (IUCN) completed a five-year Global Mammal Assessment for its IUCN Red List, which counted 5,488 species.

Taxonomy and Phylogeny

Origins

Synapsida, a clade that contains mammals and their extinct relatives, originated during the Pennsylvanian subperiod, when they split from reptilian and avian lineages. Crown group mammals evolved from earlier mammaliaforms during the Early Jurassic. The cladogram takes Mammalia to be the crown group.

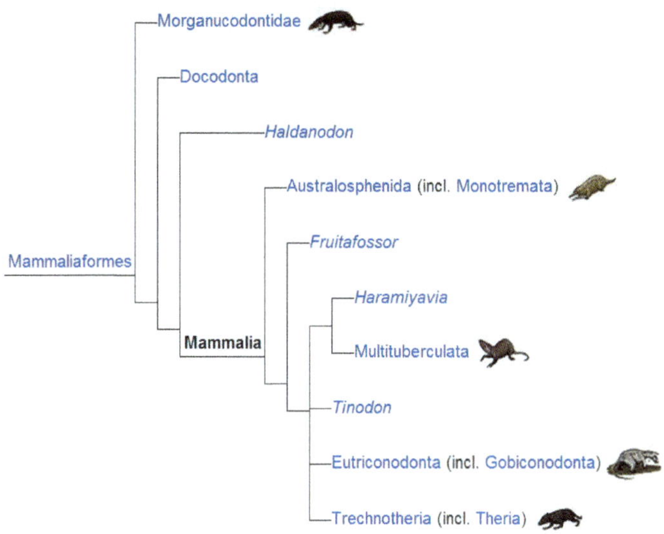

Taxonomy

The structure of Mammalia is shown in a cladogram compiled by Mikko Haaramo, based on cladograms after Rowe 1988; Luo, Crompton & Sun 2001; Luo, Cifelli & Kielan-Jaworowska 2001, Luo, Kielan-Jaworowska & Cifelli 2002, Kielan-Jaworowska, Cifelli & Luo 2004 and Luo & Wible 2005.

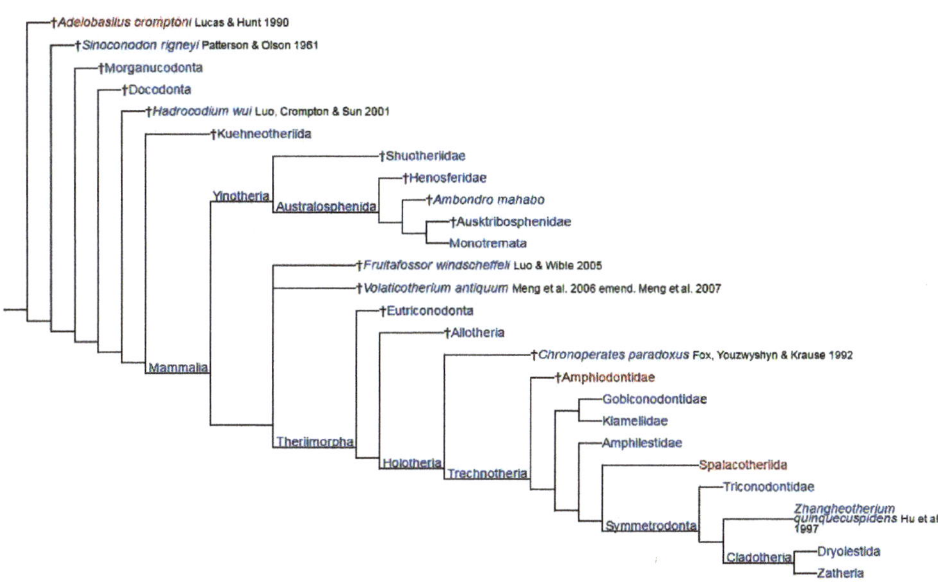

- Mammaliaformes

    - *Tricuspes* von Huene 1933 [? non-mammaliaform chiniquodontoid]
    - *Repenomamus* Li et al. 2001 {Repenomamidae Li et al. 2001} [?gobiconodont]
- Mammalia Linnaeus 1758 sensu stricto Rowe 1988 (true mammals)
    - *Reigitherium bunodontum* Bonaparte 1990 {Reigitheriidae Bonaparte 1990}
    - *Tikitherium copei* Datta 2005
- Trechnotheria
    - *Argentoconodon fariasorum* Rougier et al. 2007
    - *Austrotriconodon mckennai* Bonaparte 1986 {Austrotriconodontidae Bonaparte 1990}
    - *Dyskritodon* Sigogneau-Russell 1995
    - *Hallautherium schalchi* Clemens 1980
    - *Liaotherium gracile* Zhou, Cheng & Wang 1991
    - Jeholodentidae Luo et al. 2007
- Paraphyletic Symmetrodonta Simpson 1925
    - *Thereuodon* Sigogneau-Russell 1989 [Dryolestida, or Zatheria indet., milk tooth (Averianov, 2002)]
    - *Atlasodon* Sigogneau-Russell 1991 [Dryolestida, or Zatheria indet., milk tooth (Averianov, 2002)]
    - *Eurylambda aequicrurius*
    - *Peralestes longirostris*
    - *Tribactonodon bonfieldi* Sigogneau-Russell, Hooker & Ensom 2001

Evolution from Amniotes

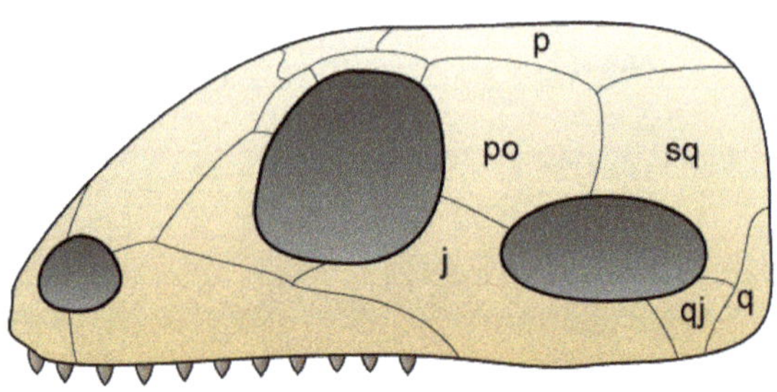

The original synapsid skull structure contains one temporal opening behind the orbitals, in a fairly low position on the skull (lower right in this image). This opening might have assisted in containing the jaw muscles of these organisms which could have increased their biting strength.

The first fully terrestrial vertebrates were amniotes. Like their amphibious tetrapod predecessors, they had lungs and limbs. Amniotic eggs, however, have internal membranes that allow the developing embryo to breathe but keep water in. Hence, amniotes can lay eggs on dry land, while amphibians generally need to lay their eggs in water.

The first amniotes apparently arose in the Late Carboniferous. They descended from earlier reptiliomorph amphibious tetrapods, which lived on land that was already inhabited by insects and other invertebrates as well as ferns, mosses and other plants. Within a few million years, two important amniote lineages became distinct: the synapsids, which would later include the common ancestor of the mammals; and the sauropsids, which now include turtles, lizards, snakes, crocodilians, dinosaurs and birds. Synapsids have a single hole (temporal fenestra) low on each side of the skull. One synapsid group, the pelycosaurs, included the largest and fiercest animals of the early Permian. Nonmammalian synapsids are sometimes called "mammal-like reptiles".

Therapsids descended from pelycosaurs in the Middle Permian, about 265 million years ago, and became the dominant land vertebrates. They differ from basal eupelycosaurs in several features of the skull and jaws, including: larger skulls and incisors which are equal in size in therapsids, but not for eupelycosaurs. The therapsid lineage leading to mammals went through a series of stages, beginning with animals that were very similar to their pelycosaur ancestors and ending with probainognathian cynodonts, some of which could easily be mistaken for mammals. Those stages were characterized by:

- The gradual development of a bony secondary palate.
- Progression towards an erect limb posture, which would increase the animals' stamina by avoiding Carrier's constraint. But this process was slow and erratic: for example, all herbivorous nonmammaliaform therapsids retained sprawling limbs (some late forms may have had semierect hind limbs); Permian carnivorous therapsids had sprawling forelimbs, and some late Permian ones also had semisprawling hindlimbs. In fact, modern monotremes still have semisprawling limbs.
- The dentary gradually became the main bone of the lower jaw which, by the Triassic, progressed towards the fully mammalian jaw (the lower consisting only of the dentary) and middle ear (which is constructed by the bones that were previously used to construct the jaws of reptiles).

First Mammals

The Permian–Triassic extinction event, which was a prolonged event due to the accumulation of several extinction pulses, ended the dominance of carnivorous therapsids. In the early Triassic, most medium to large land carnivore niches were taken over by archosaurs which, over an extended period (35 million years), came to include the crocodylomorphs, the pterosaurs and the dinosaurs; however, large cynodonts like *Trucidocynodon* and traversodontids still occupied large sized carnivorous and herbivorous niches respectively. By the Jurassic, the dinosaurs had come to dominate the large terrestrial herbivore niches as well.

The first mammals (in Kemp's sense) appeared in the Late Triassic epoch (about 225 million years ago), 40 million years after the first therapsids. They expanded out of their nocturnal insectivore niche from the mid-Jurassic onwards; The Jurassic *Castorocauda*, for example, had adaptations for swimming, digging and catching fish. Most, if not all, are thought to have remained nocturnal (the Nocturnal bottleneck), accounting for much of the typical mammalian traits. The majority of the mammal species that existed in the Mesozoic Era were multituberculates, eutriconodonts and spalacotheriids. The earliest known metatherian is *Sinodelphys*, found in 125 million-year-old Early Cretaceous shale in China's northeastern Liaoning Province. The fossil is nearly complete and includes tufts of fur and imprints of soft tissues.

Restoration of *Juramaia sinensis*, the oldest known Eutherian (160 mya)

The oldest known fossil among the Eutheria ("true beasts") is the small shrewlike *Juramaia sinensis*, or "Jurassic mother from China", dated to 160 million years ago in the late Jurassic. A later eutherian, *Eomaia*, dated to 125 million years ago in the early Cretaceous, possessed some features in common with the marsupials but not with the placentals, evidence that these features were present in the last common ancestor of the two groups but were later lost in the placental lineage. In particular, the epipubic bones extend forwards from the pelvis. These are not found in any modern placental, but they are found in marsupials, monotremes, nontherian mammals and *Ukhaatherium*, an early Cretaceous animal in the eutherian order Asioryctitheria. This also applies to the multituberculates. They are apparently an ancestral feature, which subsequently disappeared in the placental lineage. These epipubic bones seem to function by stiffening the muscles during locomotion, reducing the amount of space being presented, which placentals require to contain their fetus during gestation periods. A narrow pelvic outlet indicates that the young were very small at birth and therefore pregnancy was short, as in modern marsupials. This suggests that the placenta was a later development.

The earliest known monotreme was *Teinolophos*, which lived about 120 million years ago in Australia. Monotremes have some features which may be inherited from the original amniotes such as the same orifice to urinate, defecate and reproduce (cloaca) – as lizards and birds also do – and they lay eggs which are leathery and uncalcified.

Earliest Appearances of Features

Hadrocodium, whose fossils date from approximately 195 million years ago, in the early Jurassic, provides the first clear evidence of a jaw joint formed solely by the squamosal and dentary bones; there is no space in the jaw for the articular, a bone involved in the jaws of all early synapsids.

Foramina in the upper jaw are not indicative of whiskers, as in the red tegu (*Tupinambis rufescens*).

The earliest clear evidence of hair or fur is in fossils of *Castorocauda* and *Megaconus*, from 164 million years ago in the mid-Jurassic. In the 1950s, it was suggested that the foramina (passages) in the maxillae and premaxillae (bones in the front of the upper jaw) of cynodonts were channels which supplied blood vessels and nerves to vibrissae (whiskers) and so were evidence of hair or fur; it was soon pointed out, however, that foramina do not necessarily show that an animal had vibrissae, as the modern lizard *Tupinambis* has foramina that are almost identical to those found in the nonmammalian cynodont *Thrinaxodon*. Popular sources, nevertheless, continue to attribute whiskers to

Thrinaxodon. Studies on Permian coprolites suggest that non-mammalian synapsids of the epoch already had fur, setting the evolution of hairs possibly as far back as dicynodonts.

When endothermy first appeared in the evolution of mammals is uncertain, though it is generally agreed to have first evolved in non-mammalian therapsids. Modern monotremes have lower body temperatures and more variable metabolic rates than marsupials and placentals, but there is evidence that some of their ancestors, perhaps including ancestors of the therians, may have had body temperatures like those of modern therians. Likewise, some modern therians like afrotheres and xenarthrans have secondarily developed lower body temperatures. The evolution of erect limbs in mammals is incomplete — living and fossil monotremes have sprawling limbs. The parasagittal (nonsprawling) limb posture appeared sometime in the late Jurassic or early Cretaceous; it is found in the eutherian *Eomaia* and the metatherian *Sinodelphys*, both dated to 125 million years ago.Epipubic bones, a feature that strongly influenced the reproduction of most mammal clades, are first found in Tritylodontidae, suggesting that it is a synapomorphy between them and mammaliformes. They are omnipresent in non-placental mammaliformes, though *Megazostrodon* and *Erythrotherium* appear to have lacked them. It has been suggested that the original function of lactation (milk production) was to keep eggs moist. Much of the argument is based on monotremes, the egg-laying mammals.

Rise of the Mammals

Therian mammals took over the medium- to large-sized ecological niches in the Cenozoic, after the Cretaceous–Paleogene extinction event emptied ecological space once filled by non-avian dinosaurs and other groups of reptiles, as well as various other mammal groups. Then mammals diversified very quickly; both birds and mammals show an exponential rise in diversity. For example, the earliest known bat dates from about 50 million years ago, only 16 million years after the extinction of the dinosaurs. Molecular phylogenetic studies initially suggested that most placental orders diverged about 100 to 85 million years ago and that modern families appeared in the period from the late Eocene through the Miocene. However, no placental fossils have been found from before the end of the Cretaceous. The earliest undisputed fossils of placentals comes from the early Paleocene, after the extinction of the dinosaurs. In particular, scientists have identified an early Paleocene animal named *Protungulatum donnae* as one of the first placental mammals. however it has been reclassified as a non-placental eutherian. Recalibrations of genetic and morphological diversity rates have suggested a terminal Maastrichtian origin for placentals, and a Paleocene origin for most modern clades. The earliest known ancestor of primates is *Archicebus achilles* from around 55 million years ago. This tiny primate weighed 20–30 grams (0.7–1.1 ounce) and could fit within a human palm.

Anatomy and Morphology

Distinguishing Features

Living mammal species can be identified by the presence of sweat glands, including those that are specialized to produce milk to nourish their young. In classifying fossils, however, other features must be used, since soft tissue glands and many other features are not visible in fossils.

Many traits shared by all living mammals appeared among the earliest members of the group:

- **Jaw joint** - The dentary (the lower jaw bone, which carries the teeth) and the squamosal (a small cranial bone) meet to form the joint. In most gnathostomes, including early therapsids, the joint consists of the articular (a small bone at the back of the lower jaw) and quadrate (a small bone at the back of the upper jaw).
- **Middle ear** - In crown-group mammals, sound is carried from the eardrum by a chain of three bones, the malleus, the incus and the stapes. Ancestrally, the malleus and the incus are derived from the articular and the quadrate bones that constituted the jaw joint of early therapsids.
- **Tooth replacement** - Teeth are replaced once or (as in toothed whales and murid rodents) not at all, rather than being replaced continually throughout life.
- **Prismatic enamel** - The enamel coating on the surface of a tooth consists of prisms, solid, rod-like structures extending from the dentin to the tooth's surface.
- **Occipital condyles** - Two knobs at the base of the skull fit into the topmost neck vertebra; most other tetrapods, in contrast, have only one such knob.

For the most part, these characteristics were not present in the Triassic ancestors of the mammals. Nearly all mammal groups possess an epipubic bone, the exception being modern placentals.

Biological Systems

Raccoon lungs being inflated manually.

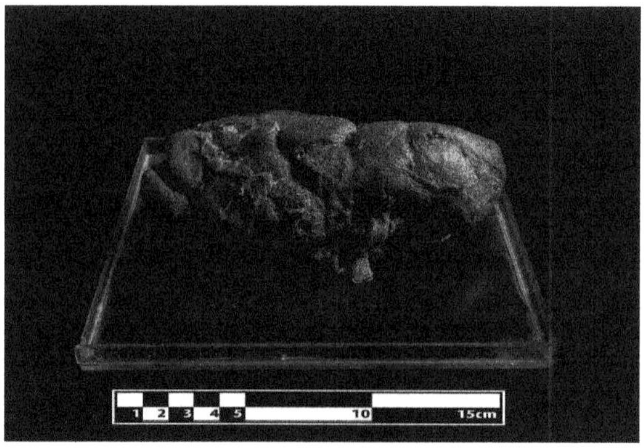

Bovine kidney

The majority of mammals have seven cervical vertebrae (bones in the neck), including bats, giraffes, whales and humans. The exceptions are the manatee and the two-toed sloth, which have just six, and the three-toed sloth which has nine cervical vertebrae. All mammalian brains possess a neocortex, a brain region unique to mammals. Placental mammals have a corpus callosum, unlike monotremes and marsupials.

The lungs of mammals are spongy and honeycombed. Breathing is mainly achieved with the diaphragm, which divides the thorax from the abdominal cavity, forming a dome convex to the thorax. Contraction of the diaphragm flattens the dome, increasing the volume of the lung cavity. Air enters through the oral and nasal cavities, and travels through the larynx, trachea and bronchi, and expands the alveoli. Relaxing the diaphragm has the opposite effect, decreasing the volume of the lung cavity, causing air to be pushed out of the lungs. During exercise, the abdominal wall contracts, increasing pressure on the diaphragm, which forces air out quicker and more forcefully. The rib cage is able to expand and contract the chest cavity through the action of other respiratory muscles. Consequently, air is sucked into or expelled out of the lungs, always moving down its pressure gradient. This type of lung is known as a bellows lung due to its resemblance to blacksmith bellows.

The integumentary system is made up of three layers: the outermost epidermis, the dermis and the hypodermis. The epidermis is typically 10 to 30 cells thick; its main function is to provide a waterproof layer. Its outermost cells are constantly lost; its bottommost cells are constantly dividing and pushing upward. The middle layer, the dermis, is 15 to 40 times thicker than the epidermis. The dermis is made up of many components, such as bony structures and blood vessels. The hypodermis is made up of adipose tissue, which stores lipids and provides cushioning and insulation. The thickness of this layer varies widely from species to species;[:97] marine mammals require a thick hypodermis (blubber) for insulation, and right whales have the thickest blubber at 20 inches (51 cm). Although other animals have features such as whiskers, feathers, setae, or cilia that superficially resemble it, no animals other than mammals have hair. It is a definitive characteristic of the class. Though some mammals have very little, careful examination reveals the characteristic, often in obscure parts of their bodies.[:61]

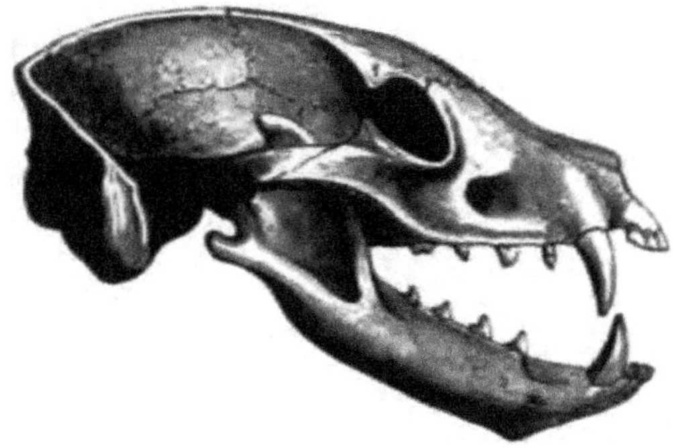

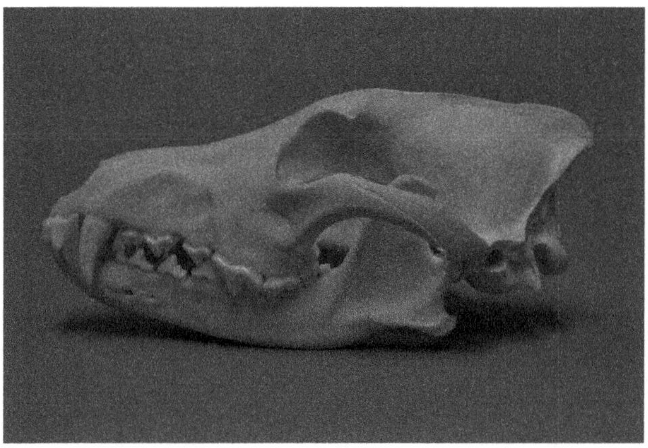

The carnassials (teeth in the very back of the mouth) of the insectivorous aardwolf (left) vs. that of a gray wolf (right) which consumes large vertebrates

Herbivores have developed a diverse range of physical structures to facilitate the consumption of plant material. To break up intact plant tissues, mammals have developed teeth structures that reflect their feeding preferences. For instance, frugivores (animals that feed primarily on fruit) and herbivores that feed on soft foliage have low-crowned teeth specialized for grinding foliage and seeds. Grazing animals that tend to eat hard, silica-rich grasses, have high-crowned teeth, which are capable of grinding tough plant tissues and do not wear down as quickly as low-crowned teeth. Most carnivorous mammals have carnassialiforme teeth (of varying length depending on diet), long canines and similar tooth replacement patterns.

The stomach of Artiodactyls is divided into four sections: the rumen, the reticulum, the omasum and the abomasum (only ruminants have a rumen). After the plant material is consumed, it is mixed with saliva in the rumen and reticulum and separates into solid and liquid material. The solids lump together to form a bolus (or cud), and is regurgitated. When the bolus enters the mouth, the fluid is squeezed out with the tongue and swallowed again. Ingested food passes to the rumen and reticulum where cellulytic microbes (bacteria, protozoa and fungi) produce cellulase, which is needed to break down the cellulose in plants. Perissodactyls, in contrast to the ruminants, store digested food that has left the stomach in an enlarged cecum, where it is fermented by bacteria. Carnivora have a simple stomach adapted to digest primarily meat, as compared to the elaborate digestive systems of herbivorous animals, which are necessary to break down tough, complex plant fibers. The caecum is either absent or short and simple, and the large intestine is not sacculated or much wider than the small intestine.

The mammalian heart has four chambers, two upper atria, the receiving chambers, and two lower ventricles, the discharging chambers. The heart has four valves, which separate its chambers and ensures blood flows in the correct direction through the heart (preventing backflow). After gas exchange in the pulmonary capillaries (blood vessels in the lungs), oxygen-rich blood returns to the left atrium via one of the four pulmonary veins. Blood flows nearly continuously back into the atrium, which acts as the receiving chamber, and from here through an opening into the left ventricle. Most blood flows passively into the heart while both the atria and ventricles are relaxed, but toward the end of the ventricular relaxation period, the left atrium will contract, pumping blood into the ventricle. The heart also requires nutrients and oxygen found in blood like other muscles, and is supplied via coronary arteries.

Coloration

A leopard's disruptively colored coat provides camouflage for this ambush predator.

Mammalian coats or pelage are colored for a variety of reasons, the major selective pressures including camouflage, sexual selection, communication and physiological processes such as temperature regulation. Camouflage is a powerful influence in a large number of mammals, as it helps to conceal individuals from predators or prey. Aposematism, warning off possible predators, is the most likely explanation of the black-and-white pelage of many mammals which are able to defend themselves, such as in the foul-smelling skunk and the powerful and aggressive honey badger. In arctic and subarctic mammals such as the arctic fox (*Alopex lagopus*), collared lemming (*Dicrostonyx groenlandicus*), stoat (*Mustela erminea*), and snowshoe hare (*Lepus americanus*), seasonal color change between brown in summer and white in winter is driven largely by camouflage. The green coloration of sloths, however, is caused by algal growths. Differences in female and male coat color may indicate nutrition and hormone levels, important in mate selection. Some primates and marsupials have shades of violet, green, or blue skin on parts of their bodies, indicating some distinct advantage in their largely arboreal habitat due to convergent evolution. Coat color is sometimes sexually dimorphic, as in many primate species.

Reproductive System

Goat kids stay with their mother until they are weaned.

Due to the presence of epipubic bones, non-placental mammals cannot expand their abdomen, being thus forced to give birth to (or lay eggs that hatch into) fetus-like larvae. Echidna "puggle"

(a) compared to various "joeys": Virginia Opossum (b), Gray short-tailed opossum (c), Eastern quoll (d), Koala (e), Brushtail possum (f) and Southern brown bandicoot (g).

Most mammals are viviparous, giving birth to live young. However, the five species of monotreme, the platypus and the four species of echidna, lay eggs. The monotremes have a sex determination system different from that of most other mammals. In particular, the sex chromosomes of a platypus are more like those of a chicken than those of a therian mammal.

The mammary glands of mammals are specialized to produce milk, the primary source of nutrition for newborns. The monotremes branched early from other mammals and do not have the nipples seen in most mammals, but they do have mammary glands. The young lick the milk from a mammary patch on the mother's belly.

Viviparous mammals are in the subclass Theria; those living today are in the marsupial and placental infraclasses. Marsupials have a short gestation period, typically shorter than its estrous cycle and gives birth to an undeveloped newborn that then undergoes further development; in many species, this takes place within a pouch-like sac, the marsupium, located in the front of the mother's abdomen. This is the plesiomorphic condition among viviparous mammals; the presence of epipubic bones in all non-placental mammals prevents the expansion of the torso needed for full pregnancy. Even non-placental eutherians probably reproduced this way. The placentals are unusual among mammals in giving birth to complete and fully developed young, usually after long gestation periods.

Endothermy

Nearly all mammals are endothermic ("warm-blooded"). Most mammals also have hair to help keep them warm. Like birds, mammals can forage or hunt in weather and climates too cold for ectothermic ("cold-blooded") reptiles and insects. Endothermy requires plenty of food energy, so mammals eat more food per unit of body weight than most reptiles. Small insectivorous mammals eat prodigious amounts for their size. A rare exception, the naked mole-rat produces little metabolic heat, so it is considered an operational poikilotherm. Birds are also endothermic, so endothermy is not unique to mammals.

Behavior

Communication and Vocalization

Many mammals communicate by vocalizing. Vocal communication serves many purposes, including in mating rituals, as warning calls, to indicate food sources, and for social purposes. Males often call during mating rituals to ward off other males and to attract females, as in the roaring of lions and red deer. The songs of the humpback whale may be signals to females; they have different dialects in different regions of the ocean. Social vocalizations include the territorial calls of gibbons, and the use of frequency in greater spear-nosed bats to distinguish between groups. The vervet monkey gives a distinct alarm call for each of at least four different predators, and the reactions of other monkeys vary according to the call. For example, if an alarm call signals a python, the monkeys climb into the trees, whereas the eagle alarm causes monkeys to seek a hiding place on the ground. Prairie dogs similarly have complex calls that signal the type, size, and speed of an approaching predator. Elephants communicate socially with a variety of sounds including snorting, screaming, trumpeting, roaring and rumbling. Some of the rumbling calls are infrasonic, below the hearing range of humans, and can be heard by other elephants up to 6 miles (9.7 km) away at still times near sunrise and sunset.

Complex communications: Vervet monkeys use at least four distinct alarm calls for different predators.

Mammals signal by a variety of means. Many give visual anti-predator signals, as when deer and gazelle stot, honestly indicating their fit condition and their ability to escape, or when white-tailed deer and other prey mammals flag with conspicuous tail markings when alarmed, informing the predator that it has been detected. Many mammals make use of scent-marking, sometimes possibly to help defend territory, but probably with a range of functions both within and between species.

Microbats and toothed whales including oceanic dolphins vocalize both socially and in echolocation.

Feeding

The insectivorous giant anteater eats some 30,000 insects per day.

To maintain a high constant body temperature is energy expensive – mammals therefore need a nutritious and plentiful diet. While the earliest mammals were probably predators, different species have since adapted to meet their dietary requirements in a variety of ways. Some eat other animals – this is a carnivorous diet (and includes insectivorous diets). Other mammals, called herbivores, eat plants. An herbivorous diet includes subtypes such as granivory (seed eating), folivory (leaf eating), frugivory (fruit eating), nectarivory (nectar eating), gummivory (gum eating) and mycophagy (fungus eating). Some mammals are coprophagous, consuming feces to absorb the nutrients not digested when the food was first ingested. An omnivore eats both prey and plants. Carnivorous mammals have a simple digestive tract because the proteins, lipids and minerals found in meat require little in the way of specialized digestion. Plants on the other hand contain complex carbohydrates, such as cellulose. The digestive tract of an herbivore is therefore host to bacteria that ferment these substances, and make them available for digestion. The bacteria are either housed in the multichambered stomach or in a large cecum.

The size of an animal is also a factor in determining diet type (Allen's rule). Since small mammals have a high ratio of heat-losing surface area to heat-generating volume, they tend to have high energy requirements and a high metabolic rate. Mammals that weigh less than about 18 oz (500 g) are mostly insectivorous because they cannot tolerate the slow, complex digestive process of an herbivore. Larger animals, on the other hand, generate more heat and less of this heat is lost. They can therefore tolerate either a slower collection process (those that prey on larger vertebrates) or a slower digestive process (herbivores). Furthermore, mammals that weigh more than 18 oz (500 g) usually cannot collect enough insects during their waking hours to sustain themselves. The only large insectivorous mammals are those that feed on huge colonies of insects (ants or termites).

Intelligence

In intelligent mammals, such as primates, the cerebrum is larger relative to the rest of the brain. Intelligence itself is not easy to define, but indications of intelligence include the ability to learn, matched with behavioral flexibility. Rats, for example, are considered to be highly intelligent, as they can learn and perform new tasks, an ability that may be important when they first colonize a fresh habitat. In some mammals, food gathering appears to be related to intelligence: a deer feeding on plants has a brain smaller than a cat, which must think to outwit its prey.

A bonobo fishing for termites with a stick

Tool use by animals may indicate different levels of learning and cognition. The sea otter uses rocks as essential and regular parts of its foraging behaviour (smashing abalone from rocks or breaking open shells), with some populations spending 21% of their time making tools. Other tool use, such as chimpanzees using twigs to "fish" for termites, may be developed by watching others use tools and may even be a true example of animal teaching. Tools may even be used in solving puzzles in which the animal appears to experience a «Eureka moment". Other mammals that do not use tools, such as dogs, can also experience a Eureka moment.

Brain size was previously considered a major indicator of the intelligence of an animal. Since most of the brain is used for maintaining bodily functions, greater ratios of brain to body mass may increase the amount of brain mass available for more complex cognitive tasks. Allometric analysis indicates that mammalian brain size scales at approximately the ⅔ or ¾ exponent of the body mass. Comparison of a particular animal's brain size with the expected brain size based on such allometric analysis provides an encephalisation quotient that can be used as another indication of animal intelligence. Sperm whales have the largest brain mass of any animal on earth, averaging 8,000 cubic centimetres (490 in^3) and 7.8 kilograms (17 lb) in mature males.

Self-awareness appears to be a sign of abstract thinking. Self-awareness, although not well-defined, is believed to be a precursor to more advanced processes such as metacognitive reasoning. The traditional method for measuring this is the mirror test, which determines if an animals possesses the ability of self-recognition. Mammals that have ‹passed' the mirror test include Asian elephants (some pass, some do not);chimpanzees; bonobos; orangutans; humans, from 18 months (mirror stage); bottlenose dolphins[a] killer whales; and false killer whales.

Social Structure

Female elephants live in stable groups, along with their offspring.

Eusociality is the highest level of social organization. These societies have an overlap of adult generations, the division of reproductive labor and cooperative caring of young. Usually insects, such as bees, ants and termites, have eusocial behavior, but it is demonstrated in two rodent species: the naked mole-rat and the Damaraland mole-rat.

Presociality is when animals exhibit more than just sexual interactions with members of the same species, but fall short of qualifying as eusocial. That is, presocial animals can display communal living, cooperative care of young, or primitive division of reproductive labor, but they do not display all of the three essential traits of eusocial animals. Humans and some species of Callitrichidae (marmosets and tamarins) are unique among primates in their degree of cooperative care of young. Harry Harlow set up an experiment with rhesus monkeys, presocial primates, in 1958; the results from this study showed that social encounters are necessary in order for the young monkeys to develop both mentally and sexually.

A fission-fusion society is a society that changes frequently in its size and composition, making up a permanent social group called the "parent group". Permanent social networks consist of all individual members of a community and often varies to track changes in their environment. In a fission–fusion society, the main parent group can fracture (fission) into smaller stable subgroups or individuals to adapt to environmental or social circumstances. For example, a number of males may break off from the main group in order to hunt or forage for food during the day, but at night they may return to join (fusion) the primary group to share food and partake in other activities. Many mammals exhibit this, such as primates (for example orangutans and spider monkeys), elephants, spotted hyenas, lions, and dolphins.

Solitary animals defend a territory and avoid social interactions with the members of its species, except during breeding season. This is to avoid resource competition, as two individuals of the same species would occupy the same niche, and to prevent depletion of food. A solitary animal, while foraging, can also be less conspicuous to predators or prey.

Red kangaroos "boxing" for dominance

In a hierarchy, individuals are either dominant or submissive. A despotic hierarchy is where one individual is dominant while the others are submissive, as in wolves and lemurs, and a pecking order is a linear ranking of individuals where there is a top individual and a bottom individual. Pecking orders may also be ranked by sex, where the lowest individual of a sex has a higher ranking than the top individual of the other sex, as in hyenas. Dominant individuals, or alphas, have a high chance of reproductive success, especially in harems where one or a few males (resident males) have exclusive breeding rights to females in a group. Non-resident males can also be accepted in harems, but some species, such as the common vampire bat (*Desmodus rotundus*), may be more strict.

When two animals mate, they both share an interest in the success of the offspring, though often to different extremes, unless the male and female are perfectly monogamous, meaning that they mate for life and take no other partners (even after the original mate's death), as with wolves, Eurasian beavers, and otters. The amount of parental care will vary. There are three types of polygamy: either one or multiple dominant males have with breeding rights (polygyny), multiple males that females mate with (polyandry), or multiple males have exclusive relations with multiple females (polygynandry). It is much more common for polygynous mating to happen, which, excluding leks, are estimated to occur in up to 90% of mammals. Lek mating occurs in harems, wherein one or a few males protect their harem of females from other males who would otherwise mate with the females, as in elephant seals; or males congregate around females and try to attract them with various courtship displays and vocalizations, as in harbor seals.

Locomotion

Terrestrial

Running gait. Photographs by Eadweard Muybridge, 1887

Most vertebrates—the amphibians, the reptiles and some mammals such as humans and bears—are plantigrade, walking on the whole of the underside of the foot. Many mammals, such as cats and dogs are digitigrade, walking on their toes, the greater stride length allowing more speed. Digitigrade mammals are also often adept at quiet movement. Some animals such as horses are unguligrade, walking on the tips of their toes. This even further increases their stride length and thus their speed. A few mammals, namely the great apes, are also known to walk on their knuckles, at least for their front legs. Giant anteaters and platypuses are also knuckle-walkers.

Animals will use different gaits for different speeds, terrain and situations. For example, horses show four natural gaits, the slowest horse gait is the walk, then there are three faster gaits which, from slowest to fastest, are the trot, the canter and the gallop. Animals may also have unusual gaits that are used occasionally, such as for moving sideways or backwards. For example, the main human gaits are bipedal walking and running, but they employ many other gaits occasionally, including a four-legged crawl in tight spaces. Mammals show a vast range of gaits, the order that they place and lift their appendages in locomotion. Gaits can be grouped into categories according to their patterns of support sequence. For quadrupeds, there are three main categories: walking gaits, running gaits and leaping gaits. Walking is the most common gait, where some feet are on the ground at any given time, and found in almost all legged animals. Running is considered to occur when at some points in the stride all feet are off the ground in a moment of suspension.

Arboreal

Arboreal animals frequently have elongated limbs that help them cross gaps, reach fruit or other resources, test the firmness of support ahead and, in some cases, to brachiate.

Many arboreal species, such as tree porcupines, silky anteaters, spider monkeys and possums, use prehensile tails to grasp branches. In the spider monkey, the tip of the tail has either a bare patch or adhesive pad, which provides increased friction.

Gibbons are very good brachiators because their elongated limbs enable them to easily swing and grasp on to branches.

Claws can be used to interact with rough substrates and re-orient the direction of forces the animal applies. This is what allows squirrels to climb tree trunks that are so large to be essentially flat from the perspective of such a small animal. However, claws can interfere with an animal's ability to grasp very small branches, as they may wrap too far around and prick the animal's own paw. Frictional gripping is used by primates, relying upon hairless fingertips. Squeezing the branch between the fingertips generates frictional force that holds the animal's hand to the branch. However, this type of grip depends upon the angle of the frictional force, thus upon the diameter of the branch, with larger branches resulting in reduced gripping ability. To control descent, especially down large diameter branches, some arboreal animals such as squirrels have evolved highly mobile ankle joints that permit rotating the foot into a 'reversed' posture. This allows the claws to hook into the rough surface of the bark, opposing the force of gravity. Small size provides many advantages to arboreal species: such as increasing the relative size of branches to the animal, lower center of mass, increased stability, lower mass (allowing movement on smaller branches) and the ability to move through more cluttered habitat. Size relating to weight affects gliding animals such as the sugar glider. Some species of primate, bat and all species of sloth achieve passive stability by hanging beneath the branch. Both pitching and tipping become irrelevant, as the only method of failure would be losing their grip.

Aerial

Slow-motion and normal speed of Egyptian fruit bats flying

Bats are the only mammals that can truly fly. They fly through the air at a constant speed by moving their wings up and down (usually with some fore-aft movement as well). Because the animal is in motion, there is some airflow relative to its body which, combined with the velocity of the wings, generates a faster airflow moving over the wing. This generates a lift force vector pointing forwards and upwards, and a drag force vector pointing rearwards and upwards. The upwards components of these counteract gravity, keeping the body in the air, while the forward component provides thrust to counteract both the drag from the wing and from the body as a whole.

The wings of bats are much thinner and consist of more bones than that of birds, allowing bats to maneuver more accurately and fly with more lift and less drag. By folding the wings inwards towards their body on the upstroke, they use 35% less energy during flight than birds. The membranes are delicate, ripping easily; however, the tissue of the bat's membrane is able to regrow, such that small tears can heal quickly. The surface of their wings is equipped with touch-sensitive receptors on small bumps called Merkel cells, also found on human fingertips. These sensitive areas are different in bats, as each bump has a tiny hair in the center, making it even more sensitive and allowing the bat to detect and collect information about the air flowing over its wings, and to fly more efficiently by changing the shape of its wings in response.

Fossorial

Semi-fossorial wombat (left) vs. fully fossorial golden mole (right)

Fossorial creatures live in subterranean environments. Many fossorial mammals were classified under the, now obsolete, order Insectivora, such as shrews, hedgehogs and moles. Fossorial mammals have a fusiform body, thickest at the shoulders and tapering off at the tail and nose. Unable to see in the dark burrows, most have degenerated eyes, but degeneration varies between species; pocket gophers, for example, are only semi-fossorial and have very small yet functional eyes, in the fully fossorial marsupial mole the eyes are degenerated and useless, talpa moles have vestigial eyes and the cape golden mole has a layer of skin covering the eyes. External ears flaps are also very small or absent. Truly fossorial mammals have short, stout legs as strength is more important than speed to a burrowing mammal, but semi-fossorial mammals have cursorial legs. The front paws are broad and have strong claws to help in loosening dirt while excavating burrows, and the back paws have webbing, as well as claws, which aids in throwing loosened dirt backwards. Most have large incisors to prevent dirt from flying into their mouth.

Aquatic

A pod of short-beaked common dolphins swimming

Fully aquatic mammals, the cetaceans and sirenians, have lost their legs and have a tail fin to propel themselves through the water. Flipper movement is continuous. Whales swim by moving their tail fin and lower body up and down, propelling themselves through vertical movement, while their flippers are mainly used for steering. Their skeletal anatomy allows them to be fast swimmers. Most species have a dorsal fin to prevent themselves from turning upside-down in the water. The flukes of sirenians are raised up and down in long strokes to move the animal forward, and can be twisted to turn. The forelimbs are paddle-like flippers which aid in turning and slowing.

Semi-aquatic mammals, like pinnipeds, have two pairs of flippers on the front and back, the fore-flippers and hind-flippers. The elbows and ankles are enclosed within the body. Pinnipeds have several adaptions for reducing drag. In addition to their streamlined bodies, they have smooth networks of muscle bundles in their skin that may increase laminar flow and make it easier for them to slip through water. They also lack arrector pili, so their fur can be streamlined as they swim. They rely on their fore-flippers for locomotion in a wing-like manner similar to penguins and sea turtles. Fore-flipper movement is not continuous, and the animal glides between each stroke. Compared to terrestrial carnivorans, the fore-limbs are reduced in length, which gives the locomotor muscles at the shoulder and elbow joints greater mechanical advantage; the hind-flippers serve as stabilizers. Other semi-aquatic mammals include beavers, hippopotamuses, otters and platypuses. Hippos are very large semi-aquatic mammals, and their barrel-shaped bodies have graviportal skeletal structures, adapted to carrying their enormous weight, and their specific gravity allows them to sink and move along the bottom of a river.

Mammals and Humans

In Human Culture

Upper Paleolithic cave painting of a variety of large mammals, Lascaux, c. 17,300 years old

Non-human mammals play a wide variety of roles in human culture. They are the most popular of pets, with tens of millions of dogs, cats and other animals including rabbits and mice kept by families around the world. Mammals such as mammoths, horses and deer are among the earliest subjects of art, being found in Upper Paleolithic cave paintings such as at Lascaux. Major artists such as Albrecht Dürer, George Stubbs and Edwin Landseer are known for their portraits of mammals. Many species of mammals have been hunted for sport and for food; deer and wild boar are especially popular as game animals. Mammals such as horses and dogs are widely raced for sport, often combined with betting on the outcome. There is a tension between the role of animals as companions to humans, and their existence as individuals with rights of their own. Mammals further play a wide variety of roles in literature, film, mythology, and religion.

Uses and Importance

Domestic mammals form a large part of the livestock raised for meat across the world. They include (2011) around 1.4 billion cattle, 1.2 billion sheep, 1 billion domestic pigs, and (1985) over 700 million rabbits. Working domestic animals including cattle and horses have been used for work and transport from the origins of agriculture, their numbers declining with the arrival of mechanised transport and agricultural machinery. In 2004 they still provided some 80% of the power for the mainly small farms in the third world, and some 20% of the world's transport, again mainly in rural areas. In mountainous regions unsuitable for wheeled vehicles, pack animals continue to transport goods. Mammal skins provide leather for shoes, clothing and upholstery. Wool from mammals including sheep, goats and alpacas has been used for centuries for clothing. Mammals serve a major role in science as experimental animals, both in fundamental biological research, such as in genetics,and in the development of new medicines, which must be tested exhaustively to demonstrate their safety.Millions of mammals, especially mice and rats, are used in experiments each year. A knockout mouse is a genetically modified mouse with an inactivated gene, replaced or disrupted with an artificial piece of DNA. They enable the study of sequenced genes whose functions are unknown. A small percentage of the mammals are non-human primates, used in research for their similarity to humans.

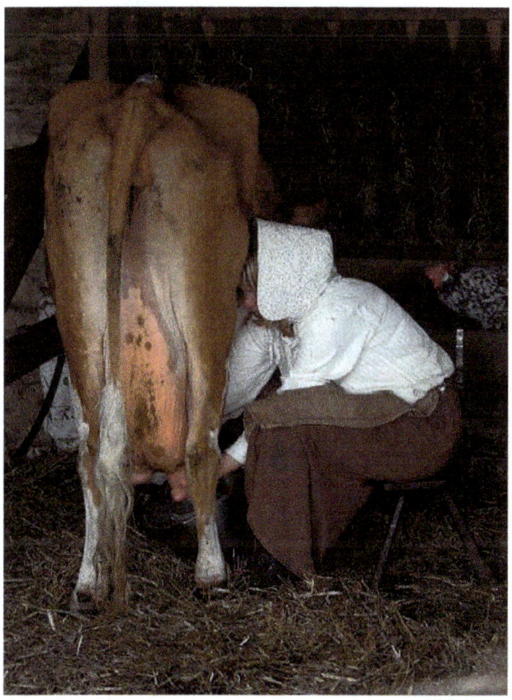

Cattle have been kept for milk for thousands of years.

Charles Darwin, Jared Diamond and others have noted the importance of domesticated mammals in the Neolithic development of agriculture and of civilization, causing farmers to replace hunter-gatherers around the world.[b] This transition from hunting and gathering to herding flocks and growing crops was a major step in human history. The new agricultural economies, based on domesticated mammals, caused "radical restructuring of human societies, worldwide alterations in biodiversity, and significant changes in the Earth's landforms and its atmosphere... momentous outcomes".

Hybrids

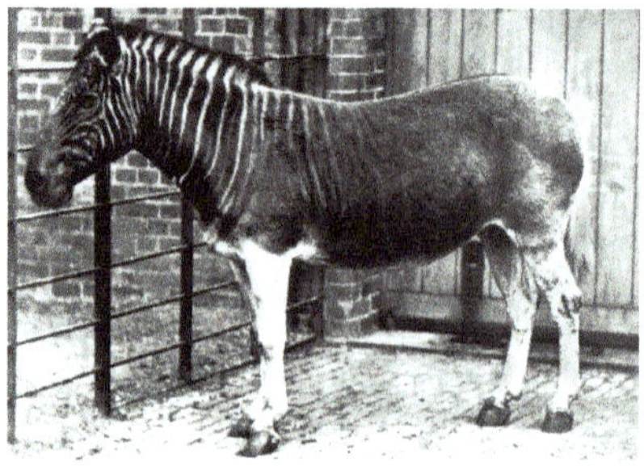

A true quagga, 1870 (left) vs. a bred-back quagga, 2014 (right)

Hybrids are offspring resulting from the breeding of two genetically distinct individuals, which usually will result in a high degree of heterozygosity, though hybrid and heterozygous are not synonymous. The deliberate or accidental hybridizing of two or more species of closely related animals through captive breeding is a human activity which has been in existence for millennia and has grown for economic purposes. Hybrids between different subspecies within a species (such as between the Bengal tiger and Siberian tiger) are known as intra-specific hybrids. Hybrids between different species within the same genus (such as between lions and tigers) are known as interspecific hybrids or crosses. Hybrids between different genera (such as between sheep and goats) are known as intergeneric hybrids. Natural hybrids will occur in hybrid zones, where two populations of species within the same genera or species living in the same or adjacent areas will interbreed with each other. Some hybrids have been recognized as species, such as the red wolf (though this is controversial).

Artificial selection, the deliberate selective breeding of domestic animals, is being used to breed back recently extinct animals in an attempt to achieve an animal breed with a phenotype that resembles that extinct wildtype ancestor. A breeding-back (intraspecific) hybrid may be very similar to the extinct wildtype in appearance, ecological niche and to some extent genetics, but the initial gene pool of that wild type is lost forever with its extinction. As a result, bred-back breeds are at best vague look-alikes of extinct wildtypes, as Heck cattle are of the aurochs.

REFERENCES

1. Vaughan, Terry A.; Ryan, James M.; Czaplewski, Nicholas J. (2013). "Classification of Mammals". Mammalogy *(6* ed.). Jones and Bartlett Learning. ISBN 978-1-284-03209-3.
2. Szalay, Frederick S. (1999). "Classification of Mammals above the Species Level: Review". Journal of Vertebrate Paleontology. 19 *(1): 191–195. doi:10.1080/02724634.1999.10011133. JSTOR 4523980.*
3. Rowe, T. (1988). "Definition, diagnosis, and origin of Mammalia" (PDF). Journal of Vertebrate Paleontology. 8 *(3): 241–264. doi:10.1080/02724634.1988.10011708.*
4. Lyell, Charles (1871). The Student's Elements of Geology. London: John Murray. p. 347. ISBN 978-1-345-18248-4.
5. Cifelli, Richard L.; Davis, Brian M. (2003). "Marsupial origins". Science. 302 *(5652): 1899–1900. doi:10.1126/science.1092272. PMID 14671280.*
6. Kemp, T. S. (2005). The Origin and Evolution of Mammals (PDF). United Kingdom: Oxford University Press. p. 3. ISBN 0-19-850760-7. OCLC 232311794.
7. Datta, P. M. (2005). "Earliest mammal with transversely expanded upper molar from the Late Triassic (Carnian) Tiki Formation, South Rewa Gondwana Basin, India". Journal of Vertebrate Paleontology. 25 *(1): 200–207.* doi:10.1671/0272-4634(2005)025[0200:EMWTEU]2.0.CO;2.
8. Luo, Zhe-Xi; Martin, Thomas (2007). "Analysis of Molar Structure and Phylogeny of Docodont Genera" (PDF). Bulletin of Carnegie Museum of Natural History. 39: 27–47. doi:10.2992/0145-9058(2007)39[27:AOMSAP]2.0.CO;2. Retrieved April 8, 2013.
9. McKenna, Malcolm C.; Bell, Susan Groag (1997). Classification of Mammals above the Species Level. New York: Columbia University Press. ISBN 0-231-11013-8. OCLC 37345734.
10. Nilsson, M. A.; Churakov, G.; Sommer, M.; van Tran, N.; Zemann, A.; Brosius, J.; Schmitz, J. (2010). "Tracking Marsupial Evolution Using Archaic Genomic Retroposon Insertions". PLoS Biology. 8 *(7): e1000436. doi:10.1371/journal.pbio.1000436. PMC 2910653.* PMID 20668664.
11. Kriegs, Jan Ole; Churakov, Gennady; Kiefmann, Martin; Jordan, Ursula; Brosius, Jürgen; Schmitz, Jürgen (2006). "Retroposed Elements as Archives for the Evolutionary History of Placental Mammals". PLoS Biology. 4 *(4): e91. doi:10.1371/journal.pbio.0040091. PMC 1395351. PMID 16515367.*
12. Nishihara, H.; Maruyama, S.; Okada, N. (2009). "Retroposon analysis and recent geological data suggest near-simultaneous divergence of the three superorders of mammals". Proceedings of the National Academy of Sciences. 106 *(13): 5235–5240. doi:10.1073/pnas.0809297106.*
13. Springer, Mark S.; Murphy, William J.; Eizirik, Eduardo; O'Brien, Stephen J. (2003). "Placental mammal diversification and the Cretaceous–Tertiary boundary". Proceedings of the National Academy of Sciences. 100 *(3): 1056–*

1061. doi:10.1073/pnas.0334222100. PMC 298725. PMID 12552136.

14. Tarver, James E.; dos Reis, Mario; Mirarab, Siavash; Moran, Raymond J.; Parker, Sean; O'Reilly, Joseph E.; King, Benjamin L.; O'Connell, Mary J.; Asher, Robert J.; Warnow, Tandy; Peterson, Kevin J.; Donoghue, Philip C. J.; Pisani, Davide (2016). "The Interrelationships of Placental Mammals and the Limits of Phylogenetic Inference". Genome Biology and Evolution. 8 *(2): 330–344. doi:10.1093/*gbe/evv261.
15. Springer, Mark S.; Meredith, Robert W.; Janecka, Jan E.; Murphy, William J. (2011). "The Historical Biogeography of Mammalia". Philosophical Transactions of the Royal Society B. 366 *(1577): 2478–2502. doi:10.1098/rstb.2011.0023. PMC 3138613. PMID 21807730.*
16. Wilson, D.E.; Reeder, D.M., eds. (2005). "Preface and introductory material". Mammal Species of the World: A Taxonomic and Geographic Reference (3rd ed.). Johns Hopkins University Press. p. xxvi. ISBN 978-0-8018-8221-0. OCLC 62265494.
17. "Mammals". The IUCN Red List of Threatened Species. IUCN. April 2010. Retrieved 23 August 2016.
18. Meng, Yuanqing Wang & Chuankui Li (2011). "Transitional mammalian middle ear from a new Cretaceous Jehol eutriconodont". Nature. 472 *(7342): 181–185. Bibcode:*2011Natur.472..181M. doi:10.1038/nature09921. PMID 21490668.
19. Haaramo, Mikko. "Mammaliaformes– mammals and near-mammals". Mikko's Phylogeny Archive.
20. Ahlberg, P. E. & Milner, A. R. (April 1994). "The Origin and Early Diversification of Tetrapods". Nature. 368 *(6471): 507–514. Bibcode:*1994Natur.368..507A. doi:10.1038/368507a0.
21. "Amniota – Palaeos". Archived from the original on 2010-12-20.
22. "Synapsida overview – Palaeos". Archived from the original on 2010-12-20.
23. Kemp, T. S. (2006). "The origin and early radiation of the therapsid mammal-like reptiles: a palaeobiological hypothesis" (PDF). Journal of Evolutionary Biology. 19 *(4): 1231–47. doi:10.1111/j.1420-9101.2005.01076.x. PMID 16780524.*
24. Bennett, A. F.; Ruben, J. A. (1986). "The metabolic and thermoregulatory status of therapsids". In Hotton III, N.; MacLean, P. D.; Roth, J. J.; Roth, E. C. The ecology and biology of mammal-like reptiles. Washington D. C.: Smithsonian Institution Press. pp. 207–218. ISBN 978-0-87474-524-5.
25. Kermack, D.M.; Kermack, K.A. (1984). The evolution of mammalian characters. Washington D.C.: Croom Helm. ISBN 0-7099-1534-9. OCLC 10710687.
26. Tanner LH, Lucas SG & Chapman MG (2004). "Assessing the record and causes of Late Triassic extinctions" (PDF). Earth-Science Reviews. 65 *(1–2): 103–139. Bibcode:*2004ESRv...65..103T. doi:10.1016/S0012-8252(03)00082-5. Archived from the original on October 25, 2007.
27. Stephen L. Brusatte. "Superiority, Competition, and Opportunism in the Evolutionary Radiation of Dinosaurs".

28. Gauthier, J.A. (1986). "Saurischian monophyly and the origin of birds". In Padian, K. (ed.). The Origin of Birds and the Evolution of Flight. Memoirs of the California Academy of Sciences. 8. San Francisco: California Academy of Sciences. pp. 1–55.

29. Sereno, P.C. (1991). "Basal archosaurs: phylogenetic relationships and functional implications". Memoirs of the Society of Vertebrate Paleontology. 2: 1–53. doi:10.2307/3889336. JSTOR 3889336.

30. MacLeod, N; Rawson, P. F.; Forey, P. L.; Banner, F. T.; Boudagher-Fadel, M. K.; Bown, P. R.; Burnett, J. A.; Chambers, P.; Culver, S.; Evans, S. E.; Jeffery, C.; Kaminski, M. A.; Lord, A. R.; Milner, A. C.; Milner, A. R.; Morris, N.; Owen, E.; Rosen, B. R.; Smith, A. B.; Taylor, P. D.; Urquhart, E.; Young, J. R. (1997). "The Cretaceous–Tertiary biotic transition". Journal of the Geological Society. 154 *(2): 265–292. doi:10.1144/gsjgs.154.2.0265.*

31. Hunt, David M.; Hankins, Mark W.; Collin, Shaun P.; Marshall, N. J. Evolution of Visual and Non-visual Pigments. London: Springer. p. 73. ISBN 978-1-4614-4354-4. OCLC 892735337.

32. Bakalar, Nicholas (2006). "Jurassic "Beaver" Found; Rewrites History of Mammals". National Geographic News. Retrieved 28 May 2016.

33. Hall, M. I.; Kamilar, J. M.; Kirk, E. C. (24 October 2012). "Eye shape and the nocturnal bottleneck of mammals". Proceedings of the Royal Society B: Biological Sciences. 279 *(1749): 4962–4968. doi:10.1098/rspb.2012.2258. PMID 23097513.*

34. Luo, Zhe-Xi (2007). "Transformation and diversification in early mammal evolution". Nature. 450 *(7172): 1011–19. Bibcode*:2007Natur.450.1011L. doi:10.1038/nature06277. PMID 18075580.

35. Pickrell, John (2003). "Oldest Marsupial Fossil Found in China". National Geographic News. Retrieved 28 May 2016.

36. Luo, Zhe-Xi; Yuan, Chong-Xi; Meng, Qing-Jin; Ji, Qiang (2011). "A Jurassic eutherian mammal and divergence of marsupials and placentals". Nature. 476 *(7361): 442–445. Bibcode*:2011Natur.476..442L. doi:10.1038/nature10291. PMID 21866158.

37. Ji, Qiang; Luo, Zhe-Xi; Yuan, Chong-Xi; Wible, John R.; Zhang, Jian-Ping; Georgi, Justin A. (2002). "The earliest known eutherian mammal". Nature. 416: 816–822. doi:10.1038/416816a. PMID 11976675.

38. M. J. Novacek; G. W. Rougier; J. R. Wible; M. C. McKenna; D. Dashzeveg & I. Horovitz (1997). "Epipubic bones in eutherian mammals from the Late Cretaceous of Mongolia". Nature. 389 *(6650): 483–486. Bibcode*:1997Natur.389..483N. doi:10.1038/39020. PMID 9333234.

39. Power, Michael L.; Schulkin, Jay (2012). "Evolution of Live Birth in Mammals". Evolution of the Human Placenta. Baltimore: Johns Hopkins University Press. p. 68. ISBN 978-1-4214-0643-5.

40. Rowe, Timothy; Rich, Thomas H.; Vickers-Rich, Patricia; Springer, Mark; Woodburne, Michael O. (2007). "The oldest platypus and its bearing on divergence

timing of the platypus and echidna clades". Proceedings of the National Academy of Sciences. 105 *(4): 1238–1242. doi:10.1073/pnas.0706385105. PMC 2234122. PMID 18216270.*

41. Grant, Tom (1995). "Reproduction". The Platypus: A Unique Mammal. Sydney: University of New South Wales. p. 55. ISBN 978-0-86840-143-0. OCLC 33842474.
42. Goldman, Armond S. (2012). "Evolution of Immune Functions of the Mammary Gland and Protection of the Infant". Breastfeeding Medicine. 7 *(3): 132–142. doi:10.1089/bfm.2012.0025.*
43. Rose, Kenneth D. (2006). The Beginning of the Age of Mammals. Baltimore: Johns Hopkins University Press. pp. 82–83. ISBN 978-0-8018-8472-6. OCLC 646769601.
44. Brink, A.S. (1955). "A study on the skeleton of Diademodon". Palaeontologia Africana. 3: 3–39.
45. Kemp, T.S. (1982). Mammal-like reptiles and the origin of mammals. London: Academic Press. p. 363. ISBN 978-0-12-404120-2. OCLC 8613180.
46. Estes, R. (1961). "Cranial anatomy of the cynodont reptile Thrinaxodon liorhinus". Bulletin of the Museum of Comparative Zoology (1253): 165–180.
47. "Thrinaxodon: The Emerging Mammal". National Geographic Daily News. February 11, 2009. Retrieved August 26, 2012.
48. Bajdek, Piotr; Qvarnström, Martin; Owocki, Krzysztof; Sulej, Tomasz; Sennikov, Andrey G.; Golubev, Valeriy K.; Niedźwiedzki, Grzegorz (2015). "Microbiota and food residues including possible evidence of pre-mammalian hair in Upper Permian coprolites from Russia". Lethaia. doi:10.1111/let.12156.
49. Botha-Brink, Jennifer; Angielczyk, Kenneth D. (2010). "Do extraordinarily high growth rates in Permo-Triassic dicynodonts (Therapsida, Anomodontia) explain their success before and after the end-Permian extinction?". *Zoological Journal of the* Linnean Society. 160 *(2): 341–365. doi:10.1111/j.1096-3642.2009.00601.x.*
50. Paul, G.S. (1988). Predatory Dinosaurs of the World. New York: Simon and Schuster. p. 464. ISBN 978-0-671-61946-6. OCLC 18350868.
51. J.M. Watson & J.A.M. Graves (1988). "Monotreme Cell-Cycles and the Evolution of Homeothermy". Australian Journal of Zoology. CSIRO. 36 *(5): 573–584.* doi:10.1071/ZO9880573.
52. McNab, Brian K. (1980). "Energetics and the limits to the temperate distribution in armadillos". Journal of Mammalogy (American Society of Mammalogists). 61 *(4): 606–627.* doi:10.2307/1380307. JSTOR 1380307.
53. Kielan−Jaworowska, Z.; Hurum, J.H.. (2006). "Limb posture in early mammals: Sprawling or parasagittal" (PDF). Acta Palaeontologica Polonica. 51 *(3): 10237–10239.*
54. Lillegraven, Jason A.; Kielan-Jaworowska, Zofia; Clemens, William A. (1979). Mesozoic Mammals: The First Two-Thirds of Mammalian History. University of California Press. p. 321. ISBN 978-0-520-03951-3. OCLC 5910695.
55. Oftedal, O.T. (2002). "The mammary gland and its origin during synapsid

evolution". Journal of Mammary Gland Biology and Neoplasia. 7 *(3): 225–252.* doi:10.1023/A:1022896515287. PMID 12751889.

56. Oftedal, O.T. (2002). "The origin of lactation as a water source for parchment-shelled eggs". Journal of Mammary Gland Biology and Neoplasia. 7 *(3): 253–266.* doi:10.1023/A:1022848632125. PMID 12751890.

57. Sahney, S., Benton, M.J. and Ferry, P.A. (2010). "Links between global taxonomic diversity, ecological diversity and the expansion of vertebrates on land" (PDF). Biology Letters. 6 *(4): 544–547. doi:10.1098/rsbl.2009.1024. PMC 2936204. PMID 20106856.*

58. Simmons, Nancy B.; Seymour, Kevin L.; Habersetzer, Jörg; Gunnell, Gregg F. (2007). "Primitive Early Eocene bat from Wyoming and the evolution of flight and echolocation". Nature. 451: 818–821. doi:10.1038/nature06549. PMID 18270539.

59. Bininda-Emonds, O.R.P.; Cardillo, M.; Jones, K.E.; Beck, Robin M. D.; Grenyer, Richard; Price, Samantha A.; Vos, Rutger A.; et al. (2007). "The delayed rise of present-day mammals" (PDF). Nature. 446 *(7135): 507–511. Bibcode*:2007Natur.446..507B. doi:10.1038/nature05634. PMID 17392779.

60. Wible, J. R.; Rogier, G. W.; Novacek, M. J.; Asher, R. J. (2007). "Cretaceous eutherians and Laurasian origin for placental mammals near the K/T boundary". Nature. 447 *(7147): 1003–06. Bibcode*:2007Natur.447.1003W. doi:10.1038/nature05854. PMID 17581585.

61. O'Leary, Maureen A.; Bloch, Jonathan I.; Flynn, John J.; Gaudin, Timothy J.; Giallombardo, Andres; Giannini, Norberto P.; Goldberg, Suzann L.; Kraatz, Brian P.; Luo, Zhe-Xi; Meng, Jin; Novacek, Michael J.; Perini, Fernando A.; Randall, Zachary S.; Rougier, Guillermo; Sargis, Eric J.; Silcox, Mary T.; Simmons, Nancy b.; Spaulding, Micelle; Velazco, Paul M.; Weksler, Marcelo; Wible, John r.; Cirranello, Andrea L.; Cirranello, Andrea L. (8 February 2013). "The Placental Mammal Ancestor and the Post–K-Pg Radiation of Placentals". Science. 339 *(6120): 662–667. Bibcode*:2013Sci...339..662O. doi:10.1126/science.1229237. PMID 23393258. Retrieved 9 February 2013.

62. Halliday, Thomas J. D.; Upchurch, Paul; Goswami, Anjali (2015). "Resolving the relationships of Paleocene placental mammals". Biological Reviews. doi:10.1111/brv.12242.

63. Halliday, Thomas John Dixon; Upchurch, Paul; Goswami, Anjali (2016). "Eutherians experienced elevated evolutionary rates in the immediate aftermath of the Cretaceous–Palaeogene mass extinction" (PDF). Proceedings of the Royal Society B. 283 *(1833): 20153026. doi:10.1098/rspb.2015.3026.*

64. Ni, Xijun; Gebo, Daniel L.; Dagosto, Marian; Meng, Jin; Tafforeau, Paul; Flynn, John J.; Beard, K. Christopher (6 June 2013). "The oldest known primate skeleton and early haplorhine evolution". Nature. 498 *(7452): 60–64. Bibcode*:2013Natur.498...60N. doi:10.1038/nature12200. PMID 23739424.

65. Romer, Sherwood A.; Parsons, Thomas S. (1977). The Vertebrate Body. Philadelphia: Holt-Saunders International. pp. 129–145. ISBN 978-0-03-910284-

5. OCLC 60007175.

66. Purves, William K.; Sadava, David E.; Orians, Gordon H.; Helle, H. C. (2001). Life: The Science of Biology (6 ed.). New York: Sinauer Associates, Inc. p. 593. ISBN 978-0-7167-3873-2. OCLC 874883911.

67. Anthwal, Neal; Joshi, Leena; Tucker, Abigail S. (2012). "Evolution of the mammalian middle ear and jaw: adaptations and novel structures". Journal of Anatomy. 221 *(1): 147–160. doi:10.1111/j.1469-7580.2012.01526.x. PMC 3552421. PMID 22686855.*

68. van Nievelt, Alexander F. H.; Smith, Kathleen K. (2005). "To replace or not to replace: the significance of reduced functional tooth replacement in marsupial and placental mammals". Paleobiology. 31 *(2): 324–346.* doi:10.1666/0094-8373(2005)031[0324:trontr]2.0.co;2.

69. Mao, Fangyuan; Wang, Yuanqing; Meng, Jin (2015). "A Systematic Study on Tooth Enamel Microstructures of Lambdopsalis bulla (Multituberculate, Mammalia) - Implications for Multituberculate Biology and Phylogeny". PLOS ONE. 10 *(5): e0128243. doi:10.1371/journal.pone.0128243. PMC 4447277. PMID 26020958.*

70. Osborn, Henry F. (1900). "Origin of the Mammalia, III. Occipital Condyles of Reptilian Tripartite Type". The American Naturalist. 34 *(408): 943–947.* doi:10.1086/277821. JSTOR 2453526.

71. Crompton, A. W.; Jenkins, Jr., F. A. (1973). "Mammals from Reptiles: A Review of Mammalian Origins". Annual Review of Earth and Planetary Sciences. 1: 131–155. doi:10.1146/annurev.ea.01.050173.001023.

72. Power, Michael L.; Schulkin, Jay (2013). The Evolution Of The Human Placenta. Baltimore: Johns Hopkins University Press. pp. 1890–1891. ISBN 978-1-4214-0643-5. OCLC 940749490.

73. Dierauf, Leslie A.; Gulland, Frances M. D. (2001). CRC Handbook of Marine Mammal Medicine: Health, Disease, and Rehabilitation (2 ed.). Boca Raton: CRC Press. p. 154. ISBN 978-1-4200-4163-7. OCLC 166505919.

74. Lui, J. H.; Hansen, D. V.; Kriegstein, A. R. (2011). "Development and Evolution of the Human Neocortex". Cell. 146 *(1): 18–36. doi:10.1016/j.cell.2011.06.030. PMC 3610574. PMID 21729779.*

75. Keeler, Clyde E. (1933). "Absence of the Corpus callosum as a Mendelizing Character in the House Mouse". Proceedings of the National Academy of Sciences of the United States of America. 19 *(6): 609–11. Bibcode:*1933PNAS...19..609K. doi:10.1073/pnas.19.6.609. JSTOR 86284. PMC 1086100. PMID 16587795.

76. Levitzky, Michael G. (2013). "Mechanics of Breathing". Pulmonary physiology (8 ed.). New York: McGraw-Hill Medical. ISBN 978-0071793131. OCLC 940633137.

77. Umesh, Kumar B. (2011). "Pulmonary Anatomy and Physiology". Handbook of Mechanical Ventilation (1 ed.). New Delhi: Jaypee Brothers Medical Publishing. p. 12. ISBN 978-93-80704-74-6. OCLC 945076700.

78. ^ Jump up to:[a] [b] [c] Feldhamer, George A.; Drickamer, Lee C.; Vessey, Stephen H.; Merritt, Joseph H.; Krajewski, Carey (2007). Mammalogy: Adaptation, Diversity, Ecology(3 ed.). Baltimore: Johns Hopkins University Press. ISBN 978-0-8018-8695-9. OCLC 124031907.

79. Tinker, Spencer W. (1988). Whales of the World. Brill Archive. p. 51. ISBN 978-0-935848-47-2.

80. Romer, A. S. (1959). The vertebrate story (4 ed.). Chicago: University of Chicago Press. ISBN 978-0-226-72490-4.

81. de Muizon, Christian; Lange-Badré, Brigitte (1997). "Carnivorous dental adaptations in tribosphenic mammals and phylogenetic reconstruction". Lethaia. 30 *(4): 353–366. doi:10.1111/j.1502-3931.1997.tb00481.x.*

82. Langer, Peter (1984). "Comparative Anatomy of the Stomach in Mammalian Herbivores". Quarterly Journal of Experimental Physiology. 69 *(3): 615–625. doi:10.1113/expphysiol.1984.sp002848. PMID 6473699.*

83. Vaughan, Terry A.; Ryan, James M.; Czaplewski, Nicholas J. (2011). "Perissodactyla". Mammalogy *(5* ed.). Jones and Bartlett. p. 322. ISBN 978-0-7637-6299-5. OCLC 437300511.

84. Flower, William H.; Lydekker, Richard (1946). An Introduction to the Study of Mammals Living and Extinct. London: Adam and Charles Black. p. 496. ISBN 978-1-110-76857-8.

85. Standring, Susan; Borley, Neil R. (2008). Gray's anatomy: the anatomical basis of clinical practice (40 ed.). London: Churchill Livingstone. pp. 960–962. ISBN 978-0-8089-2371-8. OCLC 213447727.

86. Betts, J. Gordon; Desaix, Peter; Johnson, Eddie; Johnson, Jody E.; Korol, Oksana; Kruse, Dean; Poe, Brandon; Wise, James A.; Womble, Mark; Young, Kelly A. (2013). Anatomy & physiology. Houston: Rice University Press. pp. 787–846. ISBN 978-1-938168-13-0. OCLC 898069394.

87. Caro, Tim (2005). "The Adaptive Significance of Coloration in Mammals" (PDF). BioScience. 55 *(2): 125–136.* doi:10.1641/0006-3568(2005)055[0125:tasoci]2.0.co;2.

88. Caro, Tim (February 2009). "Contrasting coloration in terrestrial mammals". Phil Trans Royal Soc B. 364 *(1516): 537–548. doi:10.1098/rstb.2008.0221. PMC 2674080. PMID 18990666.*

89. Mills, L. Scott; Zimova, Marketa; Oyler, Jared; Running, Steven; Abatzoglou, John T.; Lukacs, Paul M. (April 2013). "Camouflage mismatch in seasonal coat color due to decreased snow duration". PNAS. 110 *(8): 7360–7365. doi:10.1073/pnas.1222724110. PMC 3645584. PMID 23589881.*

90. Suutari, Milla; Majaneva, Markus; Fewer, David P.; Voirin, Bryson; Aiello, Annette; Friedl, Thomas; Chiarello, Adriano G.; Blomster, Jaanika (2010). "Molecular evidence for a diverse green algal community growing in the hair of sloths and a specific association with Trichophilus welckeri *(*Chlorophyta, Ulvophyceae)". Evolutionary Biology. 10 *(86).* doi:10.1186/1471-2148-10-86.

91. Bradley et. al, Brenda (2012). "Coat Color Variation and Pigmentation Gene Expression in Rhesus Macaques (Macaca Mulatta)" (PDF). Journal of Mammalian Evolution. 20: 263–70. doi:10.1007/s10914-012-9212-3.

92. Prum, Richard O.; Torres, Rodolfo H. (2004). "Structural colouration of mammalian skin: convergent evolution of coherently scattering dermal collagen arrays" (PDF). Journal of Experimental Biology. 207 *(12): 2157–72. doi:10.1242/jeb.00989.*

93. Plavcan, J. M. (2001). "Sexual dimorphism in primate evolution". American Journal of Physical Anthropology. 116 *(33): 25–53. doi:10.1002/ajpa.10011. PMID 11786990.*

94. Wallis M.C., Waters P.D., Delbridge M.L., Kirby P.J., Pask A.J., Grützner F., Rens W., Ferguson-Smith M.A., Graves J.A.M.; Waters; Delbridge; Kirby; Pask; Grützner; Rens; Ferguson-Smith; Graves; et al. (2007). "Sex determination in platypus and echidna: autosomal location of SOX3 confirms the absence of SRY from monotremes". Chromosome Research. 15 *(8): 949–959.* doi:10.1007/s10577-007-1185-3. PMID 18185981.

95. Marshall Graves, Jennifer A. (2008). "Weird Animal Genomes and the Evolution of Vertebrate Sex and Sex Chromosomes" (PDF). Annual Review of Genetics. 42: 568–586. doi:10.1146/annurev.genet.42.110807.091714. PMID 18983263.

96. Oftedal, O. T. (2002). "The mammary gland and its origin during synapsid evolution". Journal of Mammary Gland Biology and Neoplasia. 7 *(3): 225–252.* doi:10.1023/a:1022896515287. PMID 12751889.

97. Novacek, Michael J.; Rougier, Guillermo W.; Wible, John R.; McKenna, Malcolm C.; Dashzeveg, Demberelyin; Horovitz, Inés (1997). "Epipubic bones in eutherian mammals from the Late Cretaceous of Mongolia". Nature. 389: 483–486. doi:10.1038/39020. PMID 9333234.

98. Morgan, Sally (2005). "Mammal Behavior and Lifestyle". Mammals. Chicago: Raintree. p. 6. ISBN 978-1-4109-1050-9. OCLC 53476660.

99. Campbell, Neil A.; Reece, Jane B. (2002). Biology (6 ed.). Benjamin Cummings. p. 845. ISBN 978-080536-624-2. OCLC 47521441.

100. Buffenstein, Rochelle; Yahav, Shlomo (1991). "Is the naked mole-rat Hererocephalus glaber *an endothermic yet poikilothermic mammal?». Journal of Thermal Biology.* 16 *(4): 227–232.* doi:10.1016/0306-4565(91)90030-6.

101. Schmidt-Nielsen, Knut; Duke, James B. (1997). "Temperature Effects". Animal Physiology: Adaptation and Environment (5 ed.). Cambridge. p. 218. ISBN 978-0-521-57098-5. OCLC 35744403.

102. Seyfarth, R. M.; Cheney, D. L.; Marler, Peter (1980). "Vervet Monkey Alarm Calls: Semantic communication in a Free-Ranging Primate" (PDF). Animal Behaviour. 28 *(4): 1070–1094.* doi:10.1016/S0003-3472(80)80097-2.

103. Zuberbühler, Klause (2001). "Predator-specific alarm calls in Campbell's monkeys, Cercopithecus campbelli". Behavioral Ecology and Sociobiology. 50 *(5): 414–442.* doi:10.1007/s002650100383. JSTOR 4601985.

104. Slabbekoorn, Hans; Smith, Thomas B. (2002). "Bird song, ecology and speciation". Philosophical Transactions: Biology Sciences. 357 *(1420): 493–503. doi:10.1098/rstb.2001.1056.*

105. Bannister, John L. (2008). "Baleen Whales (Mysticetes)". In F. Perrin, William; Würsig, Bernd; Thewissen, J. G. M. Encyclopedia of Marine Mammals (2 ed.). Academic Press. pp. 80–89. ISBN 978-0-12-373553-9.

106. Norris, Scott (2002). "Creatures of Culture? Making the Case for Cultural Systems in Whales and Dolphins" (PDF). BioScience. 52 *(1): 9–14.* doi:10.1641/0006-3568(2002)052[0009:COCMTC]2.0.CO;2.

107. Boughman, Janette W. (1998). "Vocal learning by greater spear-nosed bats". Proceedings: Biological Sciences. 265 *(1392): 227–233. doi:10.1098/rspb.1998.0286.*

108. "Prairie dogs' language decoded by scientists". CBC News. 21 June 2013. Retrieved 20 May 2015.

109. Mayell, Hillary (3 March 2004). "Elephants Call Long-Distance After-Hours". National Geographic. Retrieved 15 November 2016.

110. Maynard Smith, John; Harper, David (2003). Animal Signals. Oxford Series in Ecology and Evolution. Oxford University Press. pp. 61–63. ISBN 978-0-19-852684-1. OCLC 54460090.

111. FitzGibbon, C. D.; Fanshawe, J. H. (1988). "Stotting in Thomson's gazelles: an honest signal of condition" (PDF). Behavioral Ecology and Sociobiology. 23 *(2): 69–74.* doi:10.1007/bf00299889.

112. Bildstein, Keith L. (May 1983). "Why White-Tailed Deer Flag Their Tails". The American Naturalist. 121 *(5): 709–715.* doi:10.1086/284096. JSTOR 2460873.

113. Johnson, Roger P. (August 1973). "Scent Marking in Mammals". Animal Behaviour. 21 *(3): 521–535.* doi:10.1016/S0003-3472(73)80012-0.

114. Schevill, W.E.; McBride, A.F. (1956). "Evidence for echolocation by cetaceans". Deep-Sea Research. 3 *(2): 153–154. Bibcode:*1956DSR.....3..153S. doi:10.1016/0146-6313(56)90096-x.

115. Wilson, W.; Moss, C. (2004). Thomas, J., ed. "Echolocation in Bats and Dolphins". Chicago University Press: 22. ISBN 978-0-226-79599-7. OCLC 50143737.

116. Au, Whitlow W. L. (1993). The Sonar of Dolphins. Springer-Verlag. ISBN 978-3-540-97835-0. OCLC 26158593.

117. Naugher, K. B. (2004). "Anteaters (Myrmecophagidae)". In Hutchins, M.; Kleiman, D. G.; Geist, V.; McDade, M. C. Grzimek's Animal Life Encyclopedia. 13 *(2* ed.). Gale. pp. 171–179. ISBN 978-0-7876-7750-3. OCLC 471032508.

118. Langer, Peter (1984). "Comparative Anatomy of the Stomach in Mammalian Herbivores". Quarterly Journal of Experimental Physiology. 69: 615–625. doi:10.1113/expphysiol.1984.sp002848. PMID 6473699.

119. Speaksman, J. R. (1996). "Energetics and the evolution of body size in small terrestrial mammals" (PDF). Symposia of the Zoological Society of London

(69): 69–81.

120. Don E. Wilson; David Burnie, eds. (2001). Animal: The Definitive Visual Guide to the World's Wildlife (1st ed.). DK Publishing. pp. 86–89. ISBN 978-0-7894-7764-4. OCLC 46422124.

121. Mann, Janet; Patterson, Eric M. (2013). "Tool Use by Aquatic Animals" (PDF). Philosophical Transactions of the Royal Society B. 368 *(1630): 20120424. doi:10.1098/rstb.2012.0424.*

122. Raffaele, Paul (2011). Among the Great Apes: Adventures on the Trail of Our Closest Relatives. New York: Harper. p. 83. ISBN 978-0061671-84-5. OCLC 674694369.

123. Köhler, Wolfgang (1925). The Mentality of Apes. Liveright. ISBN 978-0-87140-108-3.. OCLC 2000769.

124. McGowan, R. T.; Rehn, T.; Norling, Y.; Keeling, L. J. (2014). "Positive affect and learning: exploring the "Eureka Effect" in dogs". Animal Cognition. 17 *(13): 577–587.* doi:10.1007/s10071-013-0688-x. PMID 24096703.

125. Karbowski, Jan (2007). "Global and regional brain metabolic scaling and its functional consequences". BioMed Central Biology. 5 *(18).* doi:10.1186/1741-7007-5-18.

126. Marino, Lori (2007). "Cetacean Brains: How Aquatic Are They?". *The Anatomical Record.* 290 *(6): 694–700. doi:10.1002/ar.20530.*

127. Gallup, Jr., G. G. (1970). "Chimpanzees: Self recognition". Science. 167 *(3914): 86–87. Bibcode:*1970Sci...167...86G. doi:10.1126/science.167.3914.86. PMID 4982211.

128. Plotnik, J.M., de Waal, F.B.M. and Reiss, D. (2006). "Self-recognition in an Asian elephant" (PDF). PNAS. 103 *(45): 17053–17057. Bibcode:*2006PNAS..10317053P. doi:10.1073/pnas.0608062103.

129. S., Robert (1986). "Ontogeny of mirror behavior in two species of great apes". American Journal of Primatology. 10 *(2): 109–117. doi:10.1002/ajp.1350100202.*

130. Walraven, V., van Elsacker, L. and Verheyen, R. (1995). "Reactions of a group of pygmy chimpanzees (Pan paniscus) to their mirror images: evidence of self-recognition". Primates. 36: 145–150. doi:10.1007/bf02381922.

131. Leakey, Richard (1994). "The Origin of the Mind". The Origin Of Humankind. New York: BasicBooks. p. 150. ISBN 978-0-465-05313-1. OCLC 30739453.

132. Archer, John (1992). Ethology and Human Development. Rowman & Littlefield. pp. 215–218. ISBN 978-0-389-20996-6. OCLC 25874476.

133. Marten, K.; Psarakos, S. (1995). "Evidence of self-awareness in the bottlenose dolphin (Tursiops truncatus)". In Parker, S.T.; Mitchell, R.; Boccia, M. Self-awareness in Animals and Humans: Developmental Perspectives. Cambridge: Cambridge University Press. pp. 361–379. ISBN 978-0-521-44108-7. OCLC 28180680.

134. Delfour, F. & Marten, K. (2001). "Mirror image processing in three marine mammal species: Killer whales (Orcinus orca), false killer whales (Pseudorca crassidens)

and California sea lions (Zalophus californianus)". Behavioural Processes. 53 *(3): 181–190.* doi:10.1016/s0376-6357(01)00134-6. PMID 11334706.

135. Jarvis, J. U. M. (1981). "Eusociality in a mammal: cooperative breeding in naked mole-rat colonies". Science. 212 *(4494): 571–573. doi:10.1126/science.7209555. JSTOR 1686202.*

136. Jacobs, D.S.; et al. (1991). "The colony structure and dominance hierarchy of the Damaraland mole-rat, Cryptomys damarensis (Rodentia: Bathyergidae) from Namibia". Journal of Zoology. 224 *(4): 553–576. doi:10.1111/j.1469-7998.1991.tb03785.x.*

137. Hardy, Sarah B. (2009). Mothers and Others: The Evolutionary Origins of Mutual Understanding. Boston: Belknap Press of Harvard University Press. pp. 92–93.

138. Harlow, H. F.; Suomi, S. J. (1971). "Social Recovery by Isolation-Reared Monkeys". Proceedings of the National Academy of Sciences of the United States of America. 68 *(7): 1534–1538. doi:10.1073/pnas.68.7.1534.*

139. van Schaik, Carel P. (1999). "The Socioecology of Fission-Fusion Sociality in Orangutans". Biomedical and Life Sciences. 40 *(1): 69–86.* doi:10.1007/BF02557703.

140. Archie, Elizabeth A.; Cynthia J. Moss; Susan C. Alberts (March 2005). "The ties that bind: genetic relatedness predicts the fission and fusion of social groups in wild African elephants". Proceedings of the Royal Society B. 273: 513–522. doi:10.1098/rspb.2005.3361.

141. Smith, Jennifer E.; Sandra K. Memenis; Kay E. Holekamp (2007). "Rank-related partner choice in the fission–fusion society of the spotted hyena (Crocuta crocuta)" (PDF). Behavioral Ecology and Sociobiology. 61 *(5): 753–765.* doi:10.1007/s00265-006-0305-y.

142. Matoba, Tomoyuki; Kutsukake, Nobuyuki; Hasegawa, Toshikazu (2013). Hayward, Matt, ed. "Head Rubbing and Licking Reinforce Social Bonds in a Group of Captive African Lions, Panthera leo". PLoS ONE. 8 *(9): e73044. doi:10.1371/journal.pone.0073044. PMC 3762833. PMID 24023806.*

143. Krützen, Michael; Barré, Lynne M.; Connor, Richard C.; Mann, Janet; Sherwin, William B. (2004). "'O father: where art thou?'— Paternity assessment in an open fission–fusion society of wild bottlenose dolphins (Tursiops sp.) in Shark Bay, Western Australia". Molecular Ecology. 13 *(7): 1975–1990. doi:10.1111/j.1365-294X.2004.02192.x.*

144. Martin, Claude (1991). The Rainforests of West Africa: Ecology — Threats — Conservation (1 ed.). Springer. doi:10.1007/978-3-0348-7726-8. ISBN 978-3-0348-7726-8.

145. le Roux, Aliza; Michael I. Cherry; Lorenz Gygax (5 May 2009). "Vigilance behaviour and fitness consequences: comparing a solitary foraging and an obligate group-foraging mammal". Behavioral Ecology and Sociobiology. 63: 1097–1107. doi:10.1007/s00265-009-0762-1.

146. Palagi, Elisabetta; Norscia, Ivan (2015). Samonds, Karen E., ed. "The Season for Peace: Reconciliation in a Despotic Species (Lemur catta)". PLoS ONE.

10 *(11): e0142150. doi:10.1371/journal.pone.0142150. PMC 4646466. PMID 26569400.*

147. East, Marion L.; Hofer, Heribert (2000). "Male spotted hyenas (Crocuta crocuta) queue for status in social groups dominated by females". Behavioral Ecology. 12 *(15): 558–568. doi:10.1093/*beheco/12.5.558.
148. Samuels, A.; Silk, J. B.; Rodman, P. (1984). "Changes in the dominance rank and reproductive behavior of male bonnet macaques (Macaca radiate)". Animal Behaviour. 32: 994–1003. doi:10.1016/s0003-3472(84)80212-2.
149. Delpietro, H.A.; Russo, R.G. (2002). "Observations of the common vampire bat (Desmodus rotundus) and the hairy-legged vampire bat (Diphylla ecaudata) in captivity". Mammalian Biology. 67 *(2): 65–78.* doi:10.1078/1616-5047-00011.
150. Kleiman, Devra G. (1977). "Monogamy in Mammals". The Quarterly Review of Biology. 52 *(1): 39–69.* doi:10.1086/409721. PMID 857268.
151. Holland, B.; Rice, W. R. (1998). "Perspective: Chase-Away Sexual Selection: Antagonistic Seduction vs. Resistance" (PDF). Evolution. 52: 1–7. doi:10.2307/2410914.
152. Clutton-Brock, T. H. (1989). "Review Lecture: Mammalian Mating Systems". Proceedings of the Royal Society of London. Series B, Biological Sciences. 236 *(1285): 339–372. doi:10.1098/rspb.1989.0027. PMID 2567517.*
153. Leboeuf, J. B. (1972). "Sexual behavior in the northern elephant seal Mirounga angustirostris". Behaviour. 41 *(1): 1–26.* doi:10.1163/156853972X00167. JSTOR 4533425. PMID 5062032.
154. Boness, D. J.; Bowen, D.; Buhleier, B. M.; Marshall, G. J. (2006). "Mating tactics and mating system of an aquatic-mating pinniped: the harbor seal, Phoca vitulina" (PDF). Behavioral Ecology and Sociobiology. 61: 119–30. doi:10.1007/s00265-006-0242-9.
155. "Leg and foot". Archived from the original on 2008-04-04. Retrieved 3 August 2008.
156. Walker, Warren F.; Homberger, Dominique G. (1998). Anatomy and Dissection of the Fetal Pig (5 ed.). New York: W. H. Freeman and Company. p. 3. ISBN 978-0-7167-2637-1. OCLC 40576267.
157. Orr, CM. (2005). "Knuckle-walking anteater: a convergence test of adaptation for purported knuckle-walking features of African Hominidae". Am. J. Phys. Anthropol. 128 *(3): 639–58. doi:10.1002/ajpa.20192. PMID 15861420.*
158. Fish, FE; Frappell, PB; Baudinette, RV; MacFarlane, PM (2001). "Energetics of terrestrial locomotion of the platypus Ornithorhynchus anatinus" (PDF). The Journal of Experimental Biology. 204 *(Pt 4): 797–803. PMID 11171362.*
159. Dagg, Anne I. (1973). "Gaits in Mammals". Mammal Review. 3 *(4): 135–154. doi:10.1111/j.1365-2907.1973.tb00179.x.*
160. Roberts, Tristan D. M. (1995). Understanding Balance: The Mechanics of Posture and Locomotion. San Diego: Nelson Thornes. p. 211. ISBN 978-1-56593-416-0. OCLC 33167785.

161. Cartmill, M. (1985). "Climbing". In Hildebrand, M.; Bramble, D. M.; Liem, K. F.; Wake, D. B. Functional Vertebrate Morphology. Cambridge: Belknap Press. pp. 73–88. ISBN 978-0-674-32775-7. OCLC 11114191.

162. Vernes, Karl (2001). "Gliding Performance of the Northern Flying Squirrel (Glaucomys sabrinus) in Mature Mixed Forest of Eastern Canada" (PDF). Journal of Mammalogy. 82 *(4): 1026–1033.* doi:10.1644/1545-1542(2001)082<1026:GP OTNF>2.0.CO;2.

163. A. Barba, Lorena (October 2011). "Bats – the only flying mammals". Bio-Aerial Locomotion. Retrieved 20 May 2016.

164. «Bats In Flight Reveal Unexpected Aerodynamics". ScienceDaily. 2007. Retrieved July 12, 2016.

165. Hedenström, Anders; Johansson, L. C. (2015). "Bat flight: aerodynamics, kinematics and flight morphology" (PDF). Journal of Experimental Biology. 218: 653–663. doi:10.1242/jeb.031203.

166. "Bats save energy by drawing in wings on upstroke". ScienceDaily. 2012. Retrieved July 12, 2016.

167. Taschek, Karen (2008). Hanging with Bats: Ecobats, Vampires, and Movie Stars. Albuquerque, New Mexico: University of New Mexico Press. p. 14. ISBN 978-0-8263-4403-8. OCLC 191258477.

168. Sterbing-D'Angeloa, Susanne; Chadhab, Mohit; Chiuc, Chen; Falkc, Ben; Xianc, Wei; Barceloc, Janna; Zookd, John M.; Mossa, Cynthia F. (2011). "Bat wing sensors support flight control" (PDF). Proceedings of the National Academy of Sciences of the United States of America. 108 *(27): 11291–11296. doi:10.1073/pnas.1018740108.*

169. Shimer, H. W. (1903). "Adaptations to Aquatic, Arboreal, Fossorial and Cursorial Habits in Mammals. III. Fossorial Adaptations". The American Naturalist. 37 *(444): 819–825.* doi:10.1086/278368. JSTOR 2455381.

170. Perry, D. A. (1949). "The anatomical basis of swimming in Whales". Journal of Zoology. 119 *(1): 49–60. doi:10.1111/j.1096-3642.1949.tb00866.x.*

171. Fish, F. E.; Hui, C. A. (1991). "Dolphin swimming — a review" (PDF). Mammal Review. 21 *(4): 181–195. doi:10.1111/j.1365-2907.1991.tb00292.x.*

172. Marsh, Helene (1989). "Chapter 57: Dugongidae". Fauna of Australia (PDF). 1. Canberra: Australian Government Publications. ISBN 978-0-644-06056-1. OCLC 27492815.

173. Berta, pp. 62–64.

174. Fish, F. E. (2003). "Maneuverability by the sea lion Zalophus californianus: Turning performance of an unstable body design". Journal of Experimental Biology. 206 *(4): 667–74. doi:10.1242/jeb.00144. PMID 12517984.*

175. Riedman, M. (1990). The Pinnipeds: Seals, Sea Lions, and Walruses. University of California Press. ISBN 978-0-520-06497-3. OCLC 19511610.

176. Fish, F. E. (1996). "Transitions from drag-based to lift-based propulsion in mammalian swimming". Integrative and Comparative Biology. 36 *(6): 628–41.*

doi:10.1093/icb/36.6.628.

177. Fish, Frank E. (2000). "Biomechanics and Energetics in Aquatic and Semiaquatic Mammals: Platypus to Whale" (PDF). Physiological and Biochemical Zoology. 73 *(6): 683–698.* doi:10.1086/318108. PMID 11121343.
178. Eltringham, S. K. (1999). "Anatomy and Physiology". The Hippos. London: T & AD Poyser Ltd. p. 8. ISBN 978-0-8566-1131-5. OCLC 42274422.
179. "Hippopotamus Hippopotamus amphibius". National Geographic. Archivedfrom the original on 2014-11-25. Retrieved 30 April 2016.
180. "Animals in Healthcare Facilities" (PDF). 2012.
181. The Humane Society of the United States. "U.S. Pet Ownership Statistics". Retrieved 27 April 2012.
182. USDA. "U.S. Rabbit Industry profile" (PDF). Retrieved 10 July 2013.
183. McKie, Robin (26 May 2013). "Prehistoric cave art in the Dordogne". The Guardian. Retrieved 9 November 2016.
184. Jones, Jonathan (27 June 2014). "The top 10 animal portraits in art". The Guardian. Retrieved 24 June 2016.
185. "Deer Hunting in the United States: An Analysis of Hunter Demographics and Behavior Addendum to the 2001 National Survey of Fishing, Hunting, and Wildlife-Associated Recreation Report 2001-6". Fishery and Wildlife Service (USA). Retrieved 24 June 2016.
186. "Recreational Hog Hunting Popularity Soaring". Gramd View Outdoors. Retrieved 24 June 2016.
187. Nguyen, Jenny; Wheatley, Rick (2015). Hunting For Food: Guide to Harvesting, Field Dressing and Cooking Wild Game. F+W Media. pp. 6–77. ISBN 978-1-4403-3856-4. Chapters on hunting deer, wild hog (boar), rabbit, and squirrel.
188. "Horse racing". Archived from the original on 21 December 2013. Retrieved 6 May 2014.
189. Genders, Roy (1981). Encyclopaedia of Greyhound Racing. Pelham Books. ISBN 0-7207-1106-1.
190. Plous, S. (1993). "The Role of Animals in Human Society". Journal of Social Issues. 49 *(1): 1–9. doi:10.1111/j.1540-4560.1993.tb00906.x.*
191. Fowler, Karen Joy (26 March 2014). "Top 10 books about intelligent animals". The Guardian. Retrieved 9 November 2016.
192. Gamble, Nikki; Yates, Sally (2008). Exploring Children's Literature (2 ed.). Los Angeles: Sage. ISBN 978-1-4129-3013-0. OCLC 71285210.
193. "Books for Adults". Seal Sitters. Retrieved 9 November 2016.
194. Paterson, Jennifer (2013). "Animals in Film and Media". Oxford Bibliographies. doi:10.1093/obo/9780199791286-0044.
195. Johns, Catherine (2011). Cattle: History, Myth, Art. London: The British Museum Press. ISBN 978-0-7141-5084-0. OCLC 665137673.

196. Robert Hans van Gulik. Hayagrīva: The Mantrayānic Aspect of Horse-cult in China and Japan. Brill Archive. p. 9.
197. Grainger, Richard (24 June 2012). "Lion Depiction across Ancient and Modern Religions". ALERT. Retrieved November 6, 2016.
198. "Graphic detail Charts, maps and infographics. Counting chickens". The Economist. 27 July 2011. Retrieved November 6, 2016.
199. Cattle Today. "Breeds of Cattle at CATTLE TODAY". Cattle-today.com. Retrieved November 6, 2016.
200. Lukefahr, S.D.; Cheeke, P.R. "Rabbit project development strategies in subsistence farming systems". Food and Agriculture Organization. Retrieved November 6, 2016.
201. Pond, Wilson G. (2004). Encyclopedia of Animal Science. New York: CRC Press. pp. 248–250. ISBN 978-0-8247-5496-9. OCLC 57033325.
202. "History of Leather". Moore & Giles. Retrieved 10 November 2016.
203. Braaten, Ann W. (2005). "Wool". In Steele, Valerie. Encyclopedia of Clothing and Fashion. 3. Thomson Gale. pp. 441–443. ISBN 0-684-31394-4.
204. Quiggle, Charlotte. "Alpaca: An Ancient Luxury." Interweave Knits Fall 2000: 74-76.
205. "Genetics Research". Animal Health Trust. Retrieved November 6, 2016.
206. "Drug Development". Animal Research.info. Retrieved November 6, 2016.
207. "EU statistics show decline in animal research numbers". Speaking of Research. 2013. Retrieved November 6, 2016.
208. Helen R. Pilcher (2003). "It's a knockout". Nature. doi:10.1038/news030512-17. Retrieved November 6, 2016.
209. Y Zan et al., Production of knockout rats using ENU mutagenesis and a yeast-based screening assay, Nat. Biotechnol. (2003).Archived June 11, 2010, at the Wayback Machine.
210. «The supply and use of primates in the EU". European Biomedical Research Association. 1996. Archived from the original on 2012-01-17.
211. Carlsson, H. E.; Schapiro, S. J.; Farah, I.; Hau, J. (2004). "Use of primates in research: A global overview". American Journal of Primatology. 63 *(4): 225–237. doi:10.1002/ajp.20054. PMID 15300710.*
212. Weatherall, D., et al., (The Weatherall Committee) (2006). The use of non-human primates in research (PDF) (Report). London, UK: Academy of Medical Sciences.
213. Larson, Greger; Burger, Joachim (April 2013). "A population genetics view of animal domestication" (PDF). Trends in Genetics. 29 *(4): 197–205. doi:10.1016/j.tig.2013.01.003.*
214. Zeder, Melinda A. (August 2008). "Domestication and early agriculture in the Mediterranean Basin: Origins, diffusion, and impact". PNAS. 105 *(33): 11597–11604. doi:10.1073/pnas.0801317105. PMC 2575338. PMID 18697943.*

215. Price, E. (2008). Principles and applications of domestic animal behavior: an introductory text. Sacramento: Cambridge University Press. ISBN 978-1-84593-398-2. OCLC 226038028.

216. Taupitz, Jochen; Weschka, Marion (2009). CHIMBRIDS - Chimeras and Hybrids in Comparative European and International Research. Heidelberg: Springer. p. 13. ISBN 978-3-540-93869-9. OCLC 495479133.

217. Chambers, Steven M.; Fain, Steven R.; Fazio, Bud; Amaral, Michael (2012). "An account of the taxonomy of North American wolves from morphological and genetic analyses". North American Fauna. 77: 2. doi:10.3996/nafa.77.0001.

218. van Vuure, T. (2005). Retracing the Aurochs – History, Morphology and Ecology of an extinct wild Ox. Pensoft Publishers. ISBN 978-954-642-235-4. OCLC 940879282.

CHAPTER 6

BEHAVIORAL ECOLOGY

Chapter Contents

- **Introduction**
- **Behavioral Ecology**
- **Competing For Resources**
- **Sexual Selection**
- **Social Behaviors**
- **Altruism And Conflict In Social Insects**

Behavioral ecology, also spelled behavioural ecology, is the study of the evolutionary basis for animal behavior due to ecological pressures. Behavioral ecology emerged from ethology after Niko Tinbergen outlined four questions to address when studying animal behavior which are the proximate causes, ontogeny, survival value, and phylogeny of behavior. If an organism has a trait which provides them with a selective advantage (i.e. has an adaptive significance) in its environment, then natural selection can potentially favor it. Adaptive significance therefore refers to the beneficial qualities (such as in terms of increased survival and reproduction), any given modified trait conveys. For example, genetic differences between individuals may lead to behavioral differences, some of which in turn may drive differences in reproductive success, and ultimately over generations, the increased dominance of individuals with those favoured traits, i.e. evolution. Individuals are always in competition with others for limited resources, including food, territories, and mates. Conflict will occur between predators and prey, between rivals for mates, between siblings, mates, and even between parents and their offspring.

Penguins huddling in the Antarctic

Geese flying in aerodynamic V-formation

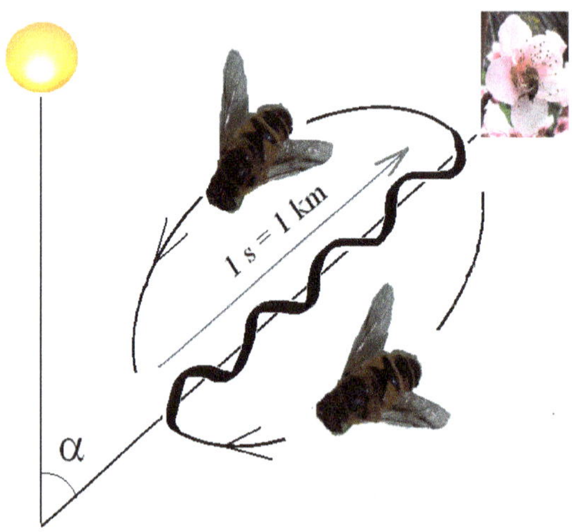

The bee waggle dance communicating information

An African elephant crossing a river

A frog with inflated vocal sac

A stotting gazelle

COMPETING FOR RESOURCES

The value of a social behavior depends in part on the social behavior of an animal's neighbors. For example, the more likely a rival male is to back down from a threat, the more value a male gets out of making the threat. The more likely, however, that a rival will attack if threatened, the less useful it is to threaten other males. When a population exhibits a number of interacting social behaviors such as this, it can evolve a stable pattern of behaviors known as an evolutionarily stable strategy (or ESS). This term, derived from economic game theory, became prominent after John Maynard Smith (1982) recognized the possible application of the concept of a Nash equilibrium to model the evolution of behavioral strategies.

Evolutionarily Stable Strategy

In short, evolutionary game theory asserts that only strategies that, when common in the population, cannot be "invaded" by any alternative (mutant) strategy will be an ESS, and thus maintained in the population. In other words, at equilibrium every player should play the best strategic response to each other. When the game is two player and symmetric each player should play the strategy which is the best response to itself.

Therefore, the ESS is considered to be the evolutionary end point subsequent to the interactions. As the fitness conveyed by a strategy is influenced by what other individuals are doing (the relative frequency of each strategy in the population), behavior can be governed not only by optimality but the frequencies of strategies adopted by others and are therefore frequency dependent (frequency dependence).

Behavioral evolution is therefore influenced by both the physical environment and interactions between other individuals.

An example of how changes in geography can make a strategy susceptible to alternative strategies is the parasitization of the African honey bee, *A. m. scutellata*.

Resource Defense

The term economic defendability was first introduced by Jerram Brown in 1964. Economic defendability states that defense of a resource have costs, such as energy expenditure or risk of injury, as well as benefits of priority access to the resource. Territorial behavior arises when benefits are greater than the costs.

Studies of the golden-winged sunbird have validated the concept of economic defendability. Comparing the energetic costs a sunbird expends in a day to the extra nectar gained by defending a territory, researchers showed that birds only became territorial when they were making a net energetic profit. When resources are at low density, the gains from excluding others may not be sufficient to pay for the cost of territorial defense. In contrast, when resource availability is high, there may be so many intruders that the defender would have no time to make use of the resources made available by defense.

Sometimes the economics of resource competition favors shared defense. An example is the feeding territories of the white wagtail. The white wagtails feed on insects washed up by the river onto the bank, which acts as a renewing food supply. If any intruders harvested their territory then the prey would quickly become depleted, but sometimes territory owners tolerate a second bird, known as a satellite. The two sharers would then move out of phase with one another, resulting in decreased feeding rate but also increased defense, illustrating advantages of group living.

Ideal Free Distribution

One of the major models used to predict the distribution of competing individuals amongst resource patches is the ideal free distribution model. Within this model, resource patches can be of variable quality, and there is no limit to the number of individuals that can occupy and extract resources from a particular patch. Competition within a particular patch means that the benefit each individual receives from exploiting a patch decreases logarithmically with increasing number of competitors sharing that resource patch. The model predicts that individuals will initially flock to higher-quality patches until the costs of crowding bring the benefits of exploiting them in line with the benefits of being the only individual on the lesser-quality resource patch. After this point has been reached, individuals will alternate between exploiting the higher-quality patches and the lower-quality patches in such a way that the average benefit for all individuals in both patches is the same. This model is *ideal* in that individuals have complete information about the quality of a resource patch and the number of individuals currently exploiting it, and *free* in that individuals are freely able to choose which resource patch to exploit.

An experiment by Manfred Malinski in 1979 demonstrated that feeding behavior in three-spined sticklebacks follows an ideal free distribution. Six fish were placed in a tank, and food items were dropped into opposite ends of the tank at different rates. The rate of food deposition at one end was set at twice that of the other end, and the fish distributed themselves with four individuals at the faster-depositing end and two individuals at the slower-depositing end. In this way, the average feeding rate was the same for all of the fish in the tank.

Mating Strategies and Tactics

As with any competition of resources, species across the animal kingdom may also engage in competitions for mating. If one considers mates or potentials mates as a resource, these sexual partners can be randomly distributed amongst resource pools within a given environment. Following the ideal free distribution model, suitors will distribute themselves amongst the potential mates in an effort to maximize their chances or the number of potential mates that they can consummate. For all competitors, males of a species in most cases, there will be variations in both the strategies and tactics used to obtain matings. Strategies generally refer to the genetically determined behaviors which can be described as conditional. Tactics refer to the subset of behaviors within a given genetic strategy. Thus it is not difficult for a great many variations in mating strategies to exist in a given environment or species.

An experiment conducted by Anthony Arak, where playback of synthetic calls from male natterjack toads was used to manipulate behavior of the males in a chorus, the difference between strategies and tactics is clear. While small and immature, male natterjack toads adopted a satellite tactic to parasitize larger males. Though large males on average still retained greater reproductive success, smaller males were able to intercept matings. When the large males of the chorus were removed, smaller males adopted a calling behavior, no longer competing against the loud calls of larger males. When smaller males got larger and their calls more competitive, then they started calling and competing directly for mates.

SEXUAL SELECTION

Mate Choice by Resources

In many sexually reproducing species, such as mammals, birds and amphibians, the females are responsible for bearing the offspring for a certain period of time, during which the males are free to mate with other available females and therefore can father many more offspring, thus continue to pass on their genes. The fundamental difference between male and female reproduction mechanisms determines the different strategies each sex employs to maximize their reproductive success. For males, their reproductive success is limited by access to females, while females are limited by their access to resources. In this sense, females can be way choosier than males because they have to bet on the resources provided by the males to ensure reproductive success.

Resources usually include nest sites, food and protection. In some cases, the males provide all of them (e.g. sedge warblers). The females dwell in their chosen males' territories for access to these resources. The males gain ownership to the territories through male-male competition that often involves physical aggression. Only the largest and strongest males manage to defend the best quality nest sites. Females choose males by inspecting the quality of different territories or by looking at some male traits that can indicate the quality of resources. Sometimes, males leave after mating. The only resource that a male provides is a nuptial gift, such as protection or food. The female can evaluate the quality of the protection or food provided by the male so as to decide whether to mate or not or how long she is willing to copulate.

Mate Choice by Genes

When males' only contribution to offspring is their sperm, females are particularly choosy. With this high level of female choice, sexual ornaments are seen in males, where the ornaments reflect the male's social status. Two hypotheses have been proposed to conceptualize the genetic benefits from female mate choice.

First, the good genes hypothesis suggests that female choice is for higher genetic quality and that this preference is favored because it increases fitness of the offspring. This includes Zahavi's handicap hypothesis and Hamilton and Zuk's host and parasite arms race. Zahavi's handicap hypothesis was proposed within the context of looking at elaborate male sexual displays. He suggested that females favor ornamented traits because they are handicaps and are indicators of the male's genetic quality. Since these ornamented traits are hazards, the male's survival must be indicative of his high genetic quality in other areas. In this way, the degree that a male expresses his sexual display indicates to the female his genetic quality. Zuk and Hamilton proposed a hypothesis after observing disease as a powerful selective pressure on a rabbit population. They suggested that sexual displays were indicators of resistance of disease on a genetic level.

Such 'choosiness' from the female individuals can be seen in wasp species too, especially among *Polistes dominula* wasps. The females tend to prefer males with smaller, more elliptically shaped spots than those with larger and more irregularly shaped spots. Those males would have reproductive superiority over males with irregular spots.

Fisher's hypothesis of runaway sexual selection suggests that female preference is genetically correlated with male traits and that the preference co-evolves with the evolution of that trait, thus the preference is under indirect selection. Fisher suggests that female preference began because the trait indicated the male's quality. The female preference spread, so that the females' offspring now benefited from the higher quality from specific trait but also greater attractiveness to mates. Eventually, the trait will only represent attractiveness to mates and no longer represent increased survival.

An example of mate choice by genes is seen in the cichlid fish *Tropheus moorii* where males provide no parental care. An experiment found that a female *T. moorii* is more likely to choose a mate with the same color morph as her own. In another experiment, females have been shown to share preferences for the same males when given two to choose from, meaning some males get to reproduce more often than others.

Sensory Bias

The sensory bias hypothesis states that the preference for a trait evolves in a non-mating context and is then exploited by one sex in order to obtain more mating opportunities. The competitive sex evolves traits that exploit a pre-existing bias that the choosy sex already possesses. This mechanism is thought to explain remarkable trait differences in closely related species because it produces a divergence in signaling systems which leads to reproductive isolation.

Sensory bias has been demonstrated in guppies, freshwater fish from Trinidad and Tobago. In this mating system, female guppies prefer to mate with males with more orange body coloration. However, outside of a mating context, both sexes prefer animate orange objects which suggests that preference originally evolved in another context, like foraging. Orange fruits are a rare treat that fall into streams where the guppies live. The ability to find these fruits quickly is an adaptive quality that has evolved outside of a mating context. Sometime after the affinity for orange objects arose, male guppies exploited this preference by incorporating large orange spots to attract females.

Another example of sensory exploitation is in the water mite *Neumania papillator*, an ambush predator which hunts copepods (small crustaceans) passing by in the water column. When hunting, *N. papillator* adopts a characteristic stance termed the 'net stance' - their first four legs are held out into the water column, with their four hind legs resting on aquatic vegetation; this allows them to detect vibrational stimuli produced by swimming prey and use this to orient towards and clutch at prey. During courtship, males actively search for females - if a male finds a female, he slowly circles around the female whilst trembling his first and second leg near her. Male leg trembling causes females (who were in the 'net stance') to orient towards often clutch the male. This did not damage the male or deter further courtship; the male then deposited spermatophores and began to vigorously fan and jerk his fourth pair of legs over the spermatophore, generating a current of water that passed over the spermatophores and towards the female. Sperm packet uptake by the female would sometimes follow. Heather Proctor hypothesised that the vibrations trembling male legs made were done to mimic the vibrations that females detect from swimming prey - this would trigger the female prey-detection responses causing females to orient and then clutch at males, mediating courtship. If this was true and males were exploiting female predation responses, then hungry females should be more receptive to male trembling – Proctor found that unfed captive females did orient and clutch at males significantly more than fed captive females did, consistent with the sensory exploitation hypothesis.

Other examples for the sensory bias mechanism include traits in auklets, wolf spiders, and manakins. Further experimental work is required to reach a fuller understanding of the prevalence and mechanisms of sensory bias.

Sexual Conflict

Sexual conflict, in some form or another, may very well be inherent in the ways most animals reproduce. Females invest more in offspring prior to mating, due to the differences in gametes in species that exhibit anisogamy, and often invest more in offspring after mating. This unequal investment leads, on one hand, to intense competition between males for mates and, on the other hand, to females choosing among males for better access to resources and good genes. Because of differences in mating goals, males and females may have very different preferred outcomes to mating.

Sexual conflict occurs whenever the preferred outcome of mating is different for the male and female. This difference, in theory, should lead to each sex evolving adaptations that bias the outcome of reproduction towards its own interests. This sexual competition leads to sexually antagonistic coevolution between males and females, resulting in what has been described as an evolutionary arms race between males and females.

Conflict Over Mating

Males' reproductive successes are often limited by access to mates, whereas females' reproductive successes are more often limited by access to resources. Thus, for a given sexual encounter, it will benefit the male to mate but the female to be choosy and resist. For example, male small tortoiseshell butterfly will compete in order to gain the best territory to mate. Another example of this conflict can be found in the Eastern carpenter bee, *Xylocopa virginica*. Males of this species are limited in reproduction primarily by access to mates, so they will claim a territory and wait for a female to pass through. Big males are, therefore, more successful in mating because they claim territories near the female nesting sites that are more sought after. Smaller males, on the other hand, will monopolize less competitive sites in foraging areas so that they may mate with reduced conflict.

Male scorpionfly

Extreme manifestations of this conflict are seen throughout nature. For example, the male *Panorpa* scorpionflies attempt to force copulation. Male scorpionflies usually acquire mates by presenting them with edible nuptial gifts in the forms of salivary secretions or dead insects. However, some males attempt to force copulation by grabbing females with a specialized abdominal organ without offering a gift. Forced copulation is costly to the female as she does not receive the food from the male and has to search for food herself (costing time and energy), while it is beneficial for the male as he does not need to find a nuptial gift.

In other cases, however, it pays for the female to gain more matings and her social mate to prevent these so as to guard paternity. For example, in many socially monogamous birds, males will follow females closely during her fertile period and attempt to chase away any other males so as to prevent extra-pair matings. The female may attempt to sneak off to achieve these extra matings. In species where males are incapable of constant guarding, the social male will often frequently copulate with the female so as to swamp rival males' sperm.

Female red junglefowl in Thailand

Sexual conflict after mating has also been shown to occur in both males and females. Males employ a diverse array of tactics to increase their success in sperm competition. These can include removing other male's sperm from females, displacing other male's sperm by flushing out prior inseminations with large amounts of their own sperm, creating copulatory plugs in females' reproductive tracts to prevent future matings with other males, spraying females with anti-aphrodisiacs to discourage other males from mating with the female, and producing sterile parasperm to protect fertile eusperm in the female's reproductive tract. Furthermore, males may control the strategic allocation of sperm, producing more sperm when females are more promiscuous. All these methods are meant to ensure that females will be more likely to produce offspring belonging to the males who uses the method.

Females also control the outcomes of matings, and there exists the possibility that females choose sperm (cryptic female choice). A dramatic example of this is the feral fowl *Gallus gallus*. In this species, females prefer to copulate with dominant males, but subordinate males can force matings. In these cases, the female is able to eject the subordinate male's sperm using cloacal contractions.

Parental Care and Family Conflicts

Parental care is the investment a parent will put into their offspring, which includes protecting and feeding the young, preparing burrows or nests, and providing eggs with yolk. There is great variation in parental care in the animal kingdom. In some species, the parents may not care for their offspring at all, while in others the parents exhibit single-parental or even bi-parental care. As with other topics in behavioral ecology, interactions within a family will involve conflicts. These conflicts can be broken down into three general types: sexual (male-female) conflict, parent-offspring conflict, and sibling conflict.

Types of Parental Care

There are many different patterns of parental care in the animal kingdom. The patterns can be explained by physiological constraints or ecological conditions, such as mating opportunities. In invertebrates, there is no parental care in most species because it is more favorable for parents to produce a large number of eggs whose fate is left to chance than to protect a few individual young. For example, female *L. figueresi* die after stocking their larvae's cells with pollen and nectar and before their larvae hatch. In birds, biparental care is the most common, because reproductive success directly depends on the parents' ability to feed their chicks. Two parents can feed twice as many young, so it is more favorable for birds to have both parents delivering food. In mammals, female-only care is the most common. This is most likely because females are internally fertilized and so are holding the young inside for a prolonged period of gestation, which provides males with the opportunity to desert. Females also feed the young through lactation after birth, so males are not required for feeding. Male parental care is only observed in species where they contribute to feeding or carrying of the young, such as in marmosets. In fish there is no parental care in 79% of bony fish. In fish with parental care, it usually limited to selecting, preparing, and defending a nest, as seen in sockeye salmon, for example. Also, parental care in fish, if any, is primarily done by males, as seen in gobies and redlip blennies. The cichlid fish *V. moorii* exhibits biparental care. In species with internal fertilization, the female is usually the one to take care of the young. In cases where fertilization is external the male becomes the main caretaker.

Familial Conflict

Familial conflict is a result of trade-offs as a function of lifetime parental investment. Parental investment was defined by Robert Trivers in 1972 as "any investment by the parent in an individual offspring that increases the offspring's chance of surviving at the cost of the parent's ability to invest in other offspring". Parental investment includes behaviors like guarding and feeding. Each parent has a limited amount of parental investment over the course of their lifetime. Investment trade-offs in offspring quality and quantity within a brood and trade offs between current and future broods leads to conflict over how much parental investment to provide and to whom parents should invest in. There are three major types of familial conflict: sexual, parent-offspring, and sibling-sibling conflict.

Sexual Conflict

There is conflict among parents as to who should provide the care as well as how much care to provide. Each parent must decide whether or not to stay and care for their offspring, or to desert their offspring. This decision is best modeled by game theoretic approaches to evolutionarily stable strategies (ESS) where the best strategy for one parent depends on the strategy adopted by the other parent. Recent research has found response matching in parents who determine how much care to invest in their offspring. Studies found that parent great tits will match their partner's increased care-giving efforts with increased provisioning rates of their own. This cued parental response is a type of behavioral negotiation between parents that leads to stabilized compensation.

Great tit

Parent-Offspring Conflict

Blackbird chicks in a nest

According to Robert Trivers's theory on relatedness, each offspring is related to itself by 1, but is only 0.5 related to their parents and siblings. Genetically, offspring are predisposed to behave in their own self-interest while parents are predisposed to behave equally to all their offspring, including both current and future ones. Offspring will selfishly attempt to take more than their fair shares of parental investment while parents will attempt to spread out their parental investment equally amongst their present young and future young. There are many examples of parent-offspring conflict in nature. One manifestation of this is asynchronous hatching in birds. A behavioral ecology hypothesis is known as Lack's brood reduction hypothesis (named after David Lack). Lack's hypothesis posits an evolutionary and ecological explanation as to why birds lay a series of eggs with an asynchronous delay leading to nestlings of mixed age and weights. According to Lack, this brood behavior is an ecological insurance that allows the larger birds to survive in poor years and all birds to survive when food is plentiful. We also see sex-ratio conflict between the queen and her workers in social hymenoptera. Because of haplodiploidy, the workers (offspring) prefer a 3:1 female to male sex allocation while the queen prefers a 1:1 sex ratio. Both the queen and the workers will attempt to bias the sex ratio in their favor. In some species, the workers gain control of the sex ratio, while in other species, like *B. terrestris*, the queen has a considerable amount of control over the colony sex ratio. Lastly, there has been recent evidence regarding genomic imprinting that is a result of parent-offspring conflict. Paternal genes in offspring will demand more maternal resources than maternal genes in the same offspring and vice versa. This has been show in imprinted genes like insulin-like growth factor-II.

Parent-Offspring Conflict Resolution

Parents need an honest signal from their offspring indicating their level of hunger or need, so that the parents can distribute resources accordingly. Offspring want to get more than their fair share of resources, so they will want to exaggerate their signals to wheedle more investment from their parents. However, this conflict is resolved by the cost of excessive begging. Not only does excessive begging attract predators, but it also retards chick growth if begging goes unrewarded. Thus, the cost of increased begging will enforce offspring honesty.

Another resolution for parent-offspring conflict is that parental provisioning and offspring demand have actually coevolved, so that there is no obvious underlying conflict. Cross-fostering experiments in great tits (*Parus major*) have shown that offspring beg more when their biological mothers are more generous. Therefore, it seems that the willingness to invest in offspring is co-adapted to offspring demand.

Sibling-Sibling Conflict

Galápagos fur seals

The lifetime parental investment is the fixed amount of parental resources available for all of a parent's young, and an offspring will want as much of it as possible. Siblings in a brood will often compete for parental resources by trying to gain more than their fair share of what their parents have to offer. There are numerous examples in nature where sibling rivalry is escalated to such an extreme that one sibling will try to kill off his other broodmates in order to maximize his own parental investment (*See Siblicide*). In the Galapagos fur seal, the second pup of a female is usually born when the first pup is still suckling. This competition for the mother's milk is especially fierce during periods of food shortage such as an El Niño year, and this usually results in the older pup directly attacking and killing the younger one.

In some bird species, sibling rivalry is also abetted by the asynchronous hatching of eggs. In the blue-footed booby, for example, the first egg in a nest is hatched four days before the second one, resulting in the elder chick having a four-day head start in growth. When the elder chick falls 20-25% below its expected weight threshold, it will attack its younger sibling and drive it from the nest.

Sibling relatedness in a brood also influences the level of sibling-sibling conflict. In a study on passerine birds, it was found that chicks begged more loudly in species with higher levels of extra-pair paternity.

Brood Parasitism

Adult reed warbler feeding a common cuckoo chick

Some animals deceive other species into providing all parental care. These brood parasites selfishly exploit their hosts' parents and host offspring. The common cuckoo is a well known example of a brood parasite. Female cuckoos lay a single egg in the nest of the host species and when the cuckoo chick hatches, it ejects all the host eggs and young. Other examples of brood parasites include honeyguides, cowbirds, and the large blue butterfly. Brood parasite offspring have many strategies to induce their host parents to invest parental care. Studies show that the common cuckoo uses vocal mimicry to reproduce the sound of multiple hungry host young to solicit more food. Other cuckoos will use visual deception with their wings to exaggerate the begging display. False gapes from brood parasite offspring cause host parents to collect more food. Another example of a brood parasite is *Phengaris* butterflies such as *Phengaris rebeli* and *Phengaris arion*, which differ from the cuckoo in that the butterflies do not oviposit directly in the nest of the host, an ant species *Myrmica schencki*.Rather, the butterfly larvae release chemicals that deceive the ants into believing that they are ant larvae, causing the ants to bring the butterfly larvae back to their own nests to feed them. Other examples of brood parasites are *Polistes sulcifer*, a paper wasp that has lost the ability to build its own nests so females lay their eggs in the nest of a host species, *Polistes dominula*, and rely on the host workers to take care of their brood, as well as *Bombus bohemicus,* a bumblebee that relies on host workers of various other *Bombus* species. Similarly, in *Eulaema meriana*, some Leucospidae wasps exploit the brood cells and nest for shelter and food from the bees.*Vespula austriaca* is another wasp in which the females force the host workers to feed and take care of the brood. In particular, *Bombus hyperboreus*, an Arctic bee species, is also classified as a brood parasite in that it attacks and enslaves other species within their subgenus, *Alpinobombus* to propagate their population.

Mating Systems

Various types of mating systems include monogamy, polygyny, polyandry, promiscuity, and polygamy. Each is differentiated by the sexual behavior between mates, such as which males mate with certain females. An influential paper by Stephen Emlen and Lewis Oring (1977)argued that two main factors of animal behavior influence the diversity of mating systems: the relative accessibility that each sex has to mates, and the parental desertion by either sex.

Mating Systems with no Male Parental Care

In a system that does not have male parental care, resource Dispersion, predation, and the effects of social living primarily influence female dispersion, which in turn influences male dispersion. Since males' primary concern is female acquisition, the males will either indirectly or directly compete for the females. In direct competition, the males are directly focused on the females. Blue-headed wrasse demonstrate the behavior in which females follow resources—such as good nest sites—and males follow the females. Conversely, species with males that exemplify indirectly competitive behavior tend towards the males' anticipation of the resources desired by females and their subsequent effort to control or acquire these resources, which helps them to achieve success with females. Grey-sided voles demonstrate indirect male competition for females. The males were experimentally observed to home in on the sites with the best food in anticipation of females settling in these areas.Males of *Euglossa imperialis*, a non-social bee species, also demonstrate indirect competitive behavior by forming aggregations of territories, which can be considered leks, to defend fragrant-rich primary territories. The purpose of these aggregations is largely only facultative, since the more suitable fragrant-rich sites there are, the more habitable territories there are to inhabit, giving females of this species a large selection of males with whom to potentially mate. Leks and choruses have also been deemed another behavior among the phenomena of male competition for females. Due to the resource-poor nature of the territories that lekking males often defend, it is difficult to categorize them as indirect competitors. Additionally, it is difficult to classify them as direct competitors seeing as they put a great deal of effort into their defense of their territories before females arrive, and upon female arrival they put for the great mating displays to attract the females to their individual sites. These observations make it difficult to determine whether female or resource dispersion primarily influences male aggregation, especially in lieu of the apparent difficulty that males may have defending resources and females in such densely populated areas. Because the reason for male aggregation into leks is unclear, five hypothesis have been proposed. These postulates propose the following as reasons for male lekking: hotspot, predation reduction, increased female attraction, hotshot males, facilitation of female choice. With all of the mating behaviors discussed, the primary factors influencing differences within and between species are ecology, social conflicts, and life history differences. In some other instances, neither direct nor indirect competition is seen. Instead, in species like the Edith's checkerspot butterfly, males' efforts are directed at acquisition of females and they exhibit indiscriminate mate location behavior, where, given the low cost of mistakes, they will blindly attempt to mate both correctly with females and incorrectly with other objects.

Mating Systems with Male Parental Care

Monogamy

Monogamy is the mating system in 90% of birds, possibly because each male and female will have a greater number of offspring if they share in raising a brood. In obligate monogamy, males feed females on the nest, or share in incubation and chick-feeding. In some species, males and females form lifelong pair bonds. Monogamy may also arise from limited opportunities for polygamy, due to strong competition among males for mates, females suffering from loss of male help, and female-female aggression.

Polygyny

In birds, polygyny occurs when males indirectly monopolize females by controlling resources. In species where males normally do not contribute much to parental care, females suffer relatively little or not at all. In other species, however, females suffer through the loss of male contribution, and the cost of having to share resources that the male controls, such as nest sites or food. In some cases, a polygynous male may control a high-quality territory so for the female, the benefits of polygyny may outweigh the costs.

Polyandry Threshold

There also seems to be a "polyandry threshold" where males may do better by agreeing to share a female instead of attempting to be in a monogamous mating system. Situations that may lead to cooperation among males include when food is scarce, and when there is intense competition for territories or females. For example, male lions sometimes form coalitions to gain control of a pride of females. In some populations of Galapagos hawks, groups of males would cooperate to defend one breeding territory. The males would share matings with the female and share paternity with the offspring.

Female Desertion and Sex Role Reversal

In birds, desertion often happens when food is abundant, so the remaining partner is better able to raise the young unaided. Desertion also occurs if there is a great chance of a parent to gain another mate, which depends on environmental and populational factors. Some birds, such as the phalaropes, have reversed sex roles where the female is larger and more brightly colored, and compete for males to incubate their clutches. In jacanas, the female is larger than the male and her territory could overlap the multiple territories of up to four males.

SOCIAL BEHAVIORS

Animals cooperate with each other in order to increase their own fitness. These altruistic, and sometimes spiteful behaviors can be explained by Hamilton's rule, which states that $rB\text{-}C > 0$ where r= relatedness, B= benefits, and C= costs.

Kin Selection

Kin selection refers to evolutionary strategies where an individual acts to favor the reproductive success of relatives, or kin, even if the action incurs some cost to the organism's own survival and ability to procreate. John Maynard Smith coined the term in 1964, although the concept was referred to by Charles Darwin who cited that helping relatives would be favored by group selection. Mathematical descriptions of kin selection were initially offered by R. A. Fisher in 1930 and J. B. S. Haldane in 1932. and 1955. W. D. Hamilton popularized the concept later, including the mathematical treatment by George Price in 1963 and 1964.

Kin selection predicts that individuals will harbor personal costs in favor of one or multiple individuals because this can maximize their genetic contribution to future generations. For example, an organism may be inclined to expend great time and energy in parental investment to rearoffspring since this future generation may be better suited for propagating genes that are highly shared between the parent and offspring.Ultimately, the initial actor performs apparent altruistic actions for kin in order to enhance its own reproductive fitness. In particular, organisms are hypothesized to act in favor of kin depending on their genetic relatedness. So, individuals will be inclined to act altruistically for siblings, grandparents, cousins and other relatives, but to differing degrees.

Inclusive Fitness

Inclusive fitness describes the component of reproductive success in both a focal individual and their relatives. Importantly, the measure embodies the sum of direct and indirect fitness and the change in their reproductive success based on the actor's behavior. That is, the effect an individual's behaviors have on: being personally better suited to reproduce offspring, and aiding descendent and non-descendent relatives in their reproductive efforts. Natural selection is predicted to push individuals to behave in ways that maximize their inclusive fitness. Studying inclusive fitness is often done using predictions from Hamilton's rule.

Kin Recognition

Genetic Cues

One possible method of kin selection is based on genetic cues that can be recognized phenotypically. Genetic recognition has been exemplified in a species that is usually not thought of as a social creature: amoebae. Social amoebae form fruiting bodies when starved for food. These amoebae preferentially formed slugs and fruiting bodies with members of their own lineage, which is clonially related. The genetic cue comes from variable lag genes, which are involved in signaling and adhesion between cells.

Kin can also be recognized a genetically determined odor, as studied in the primitively social sweat bee, *Lasioglossum zephyrum*. These bees can even recognize relatives they have never met and roughly determine relatedness. The Brazilian stingless bee *Schwarziana quadripunctata* uses a distinct combination of chemical hydrocarbons to recognize and locate kin. Each chemical odor, emitted from the organism's epicuticles, is unique and varies according to age, sex, location, and hierarchical position. Similarly, individuals of the stingless bee species *Trigona fulviventris* can distinguish kin from non-kin through recognition of a number of compounds, including hydrocarbons and fatty acids that are present in their wax and floral oils from plants used to construct their nests. In the species, *Osmia rufa,* kin selection has also been associated with mating selection. Females, specifically, will select males for mating with whom they are genetically more related to.

Environmental Cues

There are two simple rules that animals follow to determine who is kin. These rules can be exploited, but exist because they are generally successful.

The first rule is 'treat anyone in my home as kin.' This rule is readily seen in the reed warbler, a bird species that will only focus on chicks in their own nest. Interestingly, if its own kin is placed outside of the nest, a parent bird will ignore that chick. This rule can sometimes lead to odd results, especially if there is a parasitic bird that lays eggs in the reed warbler nest. For example, an adult cuckoo may sneak its egg into the nest. Once the cuckoo hatches, the reed warbler parent will feed the invading bird like its own child. Even with the risk for exploitation, the rule generally proves successful.

The second rule, named by Konrad Lorenz as 'imprinting,' states that those who you grow up with are kin. Several species exhibit this behavior, including, but not limited to the Belding's ground squirrel. Experimentation with these squirrels showed that regardless of true genetic relatedness, those that were reared together rarely fought. Further research suggests that there is partially some genetic recognition going on as well, as siblings that were raised apart were less aggressive toward one another compared to non-relatives reared apart.

Cooperation

Cooperation is broadly defined as behavior that provides a benefit to another individual that specifically evolved for that benefit. This excludes behavior that has not been expressly selected for to provide a benefit for another individual, because there are many commensal and parasitic relationships where the behavior one individual (which has evolved to benefit that individual and no others) is taken advantage of by other organisms. For cooperative behavior to be stable, it must provide a benefit to both the actor and recipient, although the benefit to the actor can take many different forms.

Within Species

Within species cooperation occurs among members of the same species. Examples of intraspecific cooperation include cooperative breeding (such as in weeper capuchins) and cooperative foraging (such as in wolves). There are also forms of cooperative defense mechanisms, such as the "fighting swarm" behavior used by the stingless bee *Tetragonula carbonaria*. Much of this behavior occurs due to kin selection. Kin selection allows cooperative behavior to evolve where the actor receives no direct benefits from the cooperation.

Cooperation (without kin selection) must evolve to provide benefits to both the actor and recipient of the behavior. This includes reciprocity, where the recipient of the cooperative behavior repays the actor at a later time. This may occur in vampire bats but it is uncommon in non-human animals. Cooperation can occur willingly between individuals when both benefit directly as well. Cooperative breeding, where one individual cares for the offspring of another, occurs in several species, including wedge-capped capuchin monkeys.

Cooperative behavior may also be enforced, where there failure to cooperate results in negative consequences. One of the best examples of this is worker policing, which occurs in social insect colonies.

Between Species

Cooperation can occur between members of different species. For interspecific cooperation to be evolutionarily stable, it must benefit individuals in both species. Examples include pistol shrimp and goby fish, nitrogen fixing microbes and legumes, ants and aphids. In ants and aphids, aphids secrete a sugary liquid called honeydew, which that ants eat. The ants provide protection to the aphids against predators, and, in some instances, raise the aphid eggs and larvae inside the ant colony. This behavior is analogous to human domestication. The genus of goby fish, *Elacatinus* also demonstrate cooperation by removing and feeding on ectoparasites of their clients. The species of wasp *Polybia rejecta* and ants *Azteca chartifex* show a cooperative behavior protecting one another's nests from predators.

Spite

Hamilton's rule can also predict spiteful behaviors between non-relatives. A spiteful behavior is one that is harmful to both the actor and to the recipient. Spiteful behavior will be favored if the actor is less related to the recipient than to the average member of the population making r negative and if rB-C is still greater than zero. Spite can also be thought of as a type of altruism because harming a non-relative, by taking his resources for example, could also benefit a relative, by allowing him access to those resources. Furthermore, certain spiteful behaviors may provide harmful short term consequences to the actor but also give long term reproductive benefits. Many behaviors that are commonly thought of as spiteful are actually better explained as being selfish, that is benefiting the actor and harming the recipient, and true spiteful behaviors are rare in the animal kingdom.

An example of spite is the sterile soldiers of the polyembryonic parasitoid wasp. A female wasp will lay a male and a female egg in a caterpillar. The eggs will divide asexually creating many genetically identical male and female larvae. Sterile soldier wasps will also develop and attack the relatively unrelated brother larvae so that the genetically identical sisters will have more access to food.

Another example is bacteria that release bacteriocins. The bacteria that releases the bacteriocin may have to die in order to do so; however most of the harm will be done to unrelated individuals who will be killed by the bacteriocin. This is because the ability to produce and release the bacteriocin is linked with immunity to it. Therefore, close relatives to the releasing cell will be less likely to die than non-relatives.

ALTRUISM AND CONFLICT IN SOCIAL INSECTS

Honeypot ant

Many insect species of the order Hymenoptera (bees, ants, wasps) are eusocial. Within the nests or hives of social insects, individuals engage in specialized tasks to ensure the survival of the colony. Dramatic examples of these specializations include changes in body morphology or unique behaviors, such as the engorged bodies of the honeypot ant *Myrmecocystus mexicanus* or the waggle dance of honey bees and a wasp species, *Vespula vulgaris*. In many, but not all social insects, reproduction is monopolized by the queen of the colony. Due to the effects of a haplodiploid mating system, in which unfertilized eggs become male drones and fertilized eggs become worker females, average relatedness values between sister workers can be higher than those seen in humans or other eutherian mammals. This has led to the suggestion that kin selection may be a driving force in the evolution of eusociality, as individuals could provide cooperative care that establishes a favorable benefit to cost ratio ($rB-c > 0$).

Naked mole-rats

However, not all social insects follow this rule. In the social wasp *Polistes dominula*, 35% of the nest mates are unrelated. In many other species, unrelated individuals only help the queen when no other options are present. In this case, subordinates work for unrelated queens even when other options may be present. No other social insect submits to unrelated queens in this way. This seemingly unfavorable behavior parallels some vertebrate systems. It is thought that this unrelated assistance is evidence of altruism in *P. dominula*. Cooperation in social organisms has numerous ecological factors that can determine the benefits and costs associated with this form of organization. One suggested benefit is a type of "life insurance" for individuals who participate in the care of the young. In this instance, individuals may have a greater likelihood of transmitting genes to the next generation when helping in a group compared to individual reproduction. Another suggested benefit is the possibility of "fortress defense", where soldier castes will threaten or attack intruders, thus protecting related individuals inside the territory. Such behaviors are seen in the snapping shrimp *Synalpheus regalis* and gall-forming aphid *Pemphigus spyrothecae*. A third ecological factor that is posited to promote eusociality is the distribution of resources: when food is sparse and concentrated in patches, eusociality is favored. Evidence supporting this third factor comes from studies of naked mole-rats and Damaraland mole-rats, which have communities containing a single pair of reproductive individuals.

Conflicts in Social Insects

Although eusociality has been shown to offer many benefits to the colony, there is also potential for conflict. Examples include the sex-ratio conflict and worker policing seen in certain species of social Hymenoptera such as *Dolichovespula media*, *Dolichovespula sylvestris, Dolichovespula norwegica* and *Vespula vulgaris.* The queen and the worker wasps either indirectly kill the laying-workers' offspring by neglecting them or directly condemn them by cannibalizing and scavenging.

The sex-ratio conflict arises from a relatedness asymmetry, which is caused by the haplodiploidy nature of Hymenoptera. For instance, workers are most related to each other because they share half of the genes from the queen and inherit all of the father's genes. Their total relatedness to each other would be $0.5+ (0.5 \times 0.5) = 0.75$. Thus, sisters are three-fourths related to each other. On the other hand, males arise from unfertilized larva, meaning they only inherit half of the queen's genes and none from the father. As a result, a female is related to her brother by 0.25, because 50% of her genes that come from her father have no chance of being shared with a brother. Her relatedness to her brother would therefore be $0.5 \times 0.5=0.25$.

According to Trivers and Hare's population-level sex-investment ratio theory, the ratio of relatedness between sexes determines the sex investment ratios. As a result, it has been observed that there is a tug-of-war between the queen and the workers, where the queen would prefer a 1:1 female to male ratio because she is equally related to her sons and daughters (r=0.5 in each case). However, the workers would prefer a 3:1 female to male ratio because they are 0.75 related to each other and only 0.25 related to their brothers. Allozyme data of a colony may indicate who wins this conflict. Conflict can also arise between workers in colonies of social insects. In some species, worker females retain their ability to mate and lay eggs. The colony's queen will be related to her sons by half of her genes and a quarter to the sons of her worker daughters. Workers, however, are related to their sons by half of their genes and to their brothers by a quarter. Thus, the queen and her worker daughters would compete for reproduction to maximize their own reproductive fitness. Worker reproduction is limited by other workers who are more related to the queen than their sisters, a situation occurring in many polyandrous hymenopteran species. Workers police the egg-laying females by engaging in oophagy or directed acts of aggression.

The Monogamy Hypothesis

The monogamy hypothesis states that the presence of monogamy in insects is crucial for eusociality to occur. This is thought to be true because of Hamilton's rule that states that rB-C>0. By having a monogamous mating system, all of the offspring have high relatedness to each other. This means that it is equally beneficial to help out a sibling, as it is to help out an offspring. If there were many fathers the relatedness of the colony would be lowered.

This monogamous mating system has been observed in insects such as termites, ants, bees and wasps. In termites the queen commits to a single male when founding a nest. In ants, bees and wasps the queens have a functional equivalent to lifetime monogamy. The male can even die before the founding of the colony. The queen can store and use the sperm from a single male throughout their lifetime, sometimes up to 30 years.

In an experiment looking at the mating of 267 hymenopteran species, the results were mapped onto a phylogeny. It was found that monogamy was the ancestral state in all the independent transitions to eusociality. This indicates that monogamy is the ancestral, likely to be crucial state for the development of eusociality. In species where queens mated with multiple mates, it was found that these were developed from lineages where sterile castes already evolved, so the multiple mating was secondary. In these cases, multiple mating is likely to be advantageous for reasons other than those important at the origin of eusociality. Most likely reasons are that a diverse worker pool attained by multiple mating by the queen increases disease resistance and may facilitate a division of labor among workers

Communication and Signalling

Communication is varied at all scales of life, from interactions between microscopic organisms to those of large groups of people. Nevertheless, the signals used in communication abide by a fundamental property: they must be a quality of the receiver that can transfer information to a receiver that is capable of interpreting the signal and modifying its behavior accordingly. Signals are distinct from cues in that evolution has selected for signalling between both parties, whereas cues are merely informative to the observer and may not have originally been used for the intended purpose. The natural world is replete with examples of signals, from the luminescent flashes of light from fireflies, to chemical signaling in red harvester ants to prominent mating displays of birds such as the Guianan cock-of-the-rock, which gather in leks, the pheromones released by the corn earworm moth, the dancing patterns of the blue-footed booby, or the alarm sound *Synoeca cyanea* make by rubbing their mandibles against their nest.

The nature of communication poses evolutionary concerns, such as the potential for deceit or manipulation on the part of the sender. In this situation, the receiver must be able to anticipate the interests of the sender and act appropriately to a given signal. Should any side gain advantage in the short term, evolution would select against the signal or the response. The conflict of interests between the sender and the receiver results in an evolutionarily stable state only if both sides can derive an overall benefit.

Although the potential benefits of deceit could be great in terms of mating success, there are several possibilities for how dishonesty is controlled, which include indices, handicaps, and common interests. Indices are reliable indicators of a desirable quality, such as overall health, fertility, or fighting ability of the organism. Handicaps, as the term suggests, place a restrictive cost on the organisms that own them, and thus lower quality competitors experience a greater relative cost compared to their higher quality counterparts. In the common interest situation, it is beneficial to both sender and receiver to communicate honestly such that the benefit of the interaction is maximized.

Signals are often honest, but there are exceptions. Prime examples of dishonest signals include the luminescent lure of the anglerfish, which is used to attract prey, or the mimicry of non-poisonous butterfly species, like the Batesian mimic *Papilio polyxenes* of the poisonous model *Battus philenor*. Although evolution should normally favor selection against the dishonest signal, in these cases it appears that the receiver would benefit more on average by accepting the signal.

REFERENCES

1. Maynard Smith, J. 1982. Evolution and the Theory of Games.
2. Brown, Jerram (June 1964). "The evolution of diversity in avian territorial systems". The Wilson Bulletin. 76 (2): 160–169. JSTOR 4159278.
3. Gill, Frank; Larry Wolf (1975). "Economics of feeding territoriality in the golden-winged sunbird". Ecology. 56 (2): 333–345. doi:10.2307/1934964. JSTOR 1934964.
4. Davies, N. B.; A. I. Houston (Feb 1981). "Owners and satellites: the economics of territory defence in the pied wagtail, Motacilla alba". Journal of Animal Ecology. 50 (1): 157–180. doi:10.2307/4038. JSTOR 4038.
5. Fretwell, Stephen D. (1972). Population in a Seasonal Environment. Princeton, NJ: Princeton University Press.
6. Milinski, Manfred (1979). "An Evolutionarily Stable Feeding Strategy in Sticklebacks". Zeitschrift für Tierpsychologie. 51 (1): 36–40. doi:10.1111/j.1439-0310.1979.tb00669.x.
7. Dominey, Wallace (1984). "Alternative Mating Tactics and Evolutionarily Stable Strategies". American Zoology. 24 (2): 385–396. doi:10.1093/icb/24.2.385.
8. Arak, Anthony (1983). "Sexual selection by male-male competition in natterjack toad choruses". Nature. 306 (5940): 261–262. Bibcode:1983Natur.306..261A.

doi:10.1038/306261a0.

9. Nicholas B. Davies; John R. Krebs; Stuart A. West (2012). An Introduction to Behavioral Ecology. West Sussex, UK: Wiley-Blackwell. pp. 193–202. ISBN 978-1-4051-1416-5.
10. Buchanan, K.L.; Catchpole, C.K. (2000). "Song as an indicator of male parental effort in the sedge warbler". Proceedings of the Royal Society. B. 267 (1441): 321–326. doi:10.1098/rspb.2000.1003.
11. Dussourd, D.E.; Harvis, C.A.; Meinwald, J.; Eisner, T. (1991). "Pheromonal advertisement of a nuptial gift by a male moth". Proceedings of the National Academy of Sciences of the United States of America. 88 (20): 9224–9227. Bibcode:1991PNAS...88.9224D. doi:10.1073/pnas.88.20.9224. PMC 52686 . PMID 1924385.
12. Ryan, Michael J.; Anne Keddy-Hector (March 1992). "Directional patterns of female mate choice and the role of sensory biases". The American Naturalist. 139: S4–S35. doi:10.1086/285303. JSTOR 2462426.
13. Salzburger, Walter; Niederstätter, Harald; Brandstätter, Anita; Berger, Burkhard; Parson, Walther; Snoeks, Jos; Sturmbauer, Christian (2006). "Colour-assortative mating among populations of Tropheus moorii, a cichlid fish from Lake Tanganyika, East Africa". Proceedings of the Royal Society B: Biological Sciences. 273 (1584): 257–66. doi:10.1098/rspb.2005.3321.
14. Steinwender, Bernd; Koblmüller, Stephan; Sefc, Kristina M. "Concordant female mate preferences in the cichlid fish Tropheus moorii". Hydrobiologia. 682: 121–130. doi:10.1007/s10750-011-0766-5.
15. Boughman, J. W. (2002). "How sensory drive can promote speciation". Trends in Ecology and Evolution. 17 (12): 571–577. doi:10.1016/S0169-5347(02)02595-8.
16. Rodd, F. H.; Hughes, K. A.; Grether, G. F.; Baril, C. T. (2002). "A possible non-sexual origin of mate preference: are male guppies mimicking fruit?". Proceedings of the Royal Society B. 7 (1490): 475–481. doi:10.1098/rspb.2001.1891.
17. Proctor, Heather C. (1991-10-01). "Courtship in the water mite Neumania papillator: males capitalize on female adaptations for predation". Animal Behaviour. 42 (4): 589–598. doi:10.1016/S0003-3472(05)80242-8.
18. Proctor, Heather C. (1992-10-01). "Sensory exploitation and the evolution of male mating behaviour: a cladistic test using water mites (Acari: Parasitengona)". Animal Behaviour. 44 (4): 745–752. doi:10.1016/S0003-3472(05)80300-8.
19. Proctor, H. C. (1992-01-01). "Effect of Food Deprivation on Mate Searching and Spermatophore Production in Male Water Mites (Acari: Unionicolidae)". Functional Ecology. 6 (6): 661–665. doi:10.2307/2389961. JSTOR 2389961.
20. Alcock, John. Animal Behaviour: A Evolutionary Approach (10th ed.). Sinauer. pp. 70–72. ISBN 9780878939664.
21. Jones, I. L.; Hunter, F. M. (1998). "Heterospecific mating preferences for a feather ornament in least auklets". Behavioral Ecology. 9 (2): 187–192. doi:10.1093/beheco/9.2.187.

22. McClinktock, W. J.; Uetz, G. W. (1996). "Female choice and pre-existing bias: Visual cues during courtship in two Schizocosawolff spiders". Animal Behaviour. 52: 167–181. doi:10.1006/anbe.1996.0162.
23. Prum, R. O. (1996). "Phylogenetic tests of alternative intersexual selection mechanisms: Trait macroevolution in a polygynous clade". The American Naturalist. 149 (4): 688–692. doi:10.1086/286014. JSTOR 2463543.
24. Fuller, R. C.; Houle, D.; Travis, J. (2005). "Sensory bias as an explanation for the evolution of mate preferences". American Naturalist. 166 (4): 437–446. doi:10.1086/444443. PMID 16224700.
25. Parker, G. A. (2006). "Sexual conflict over mating and fertilization: An overview". Philosophical Transactions of the Royal Society B. 361 (1466): 235–59. doi:10.1098/rstb.2005.1785. PMC 1569603 . PMID 16612884.
26. Davies N, Krebs J, and West S. (2012). An Introduction to Behavioral Ecology, 4th Ed. Wiley-Blackwell; Oxford: page 209-220.
27. Parker, G. (1979). "Sexual selection and sexual conflict." In: Sexual Selection and Reproductive Competition in Insects (eds. M.S. Blum and N.A. Blum). Academic Press, New York: pp123-166.
28. Chapman, T.; et al. (2003). "Sexual Selection". Trends in Ecology and Evolution. 18: 41–47. doi:10.1016/s0169-5347(02)00004-6.
29. Baker, R. R. (1972). "Territorial behaviour of the Nymphalid butterflies, Aglais urticae (L.) and Inachis io (L.)". Journal of Animal Ecology. 41 (2): 453–469. doi:10.2307/3480.
30. Skandalis, Dimitri A.; Tattersall, Glenn J.; Prager, Sean; Richards, Miriam H. (2009). "Body Size and Shape of the Large Carpenter Bee, Xylocopa virginica (L.) (Hymenoptera: Apidae)". Journal of the Kansas Entomological Society. 82 (1): 30–42. doi:10.2317/JKES711.05.1.
31. Thornhill, R. (1980). "Rape in Panorpa scorpionflies and a general rape hypothesis". Animal Behaviour. 28: 52–59. doi:10.1016/s0003-3472(80)80007-8.
32. Birkhead, T. and Moller, A. (1992). Sperm Competition in Birds: Evolutionary Causes and Consequences. Academic Press, London.
33. Pizzari, T.; Birkhead, T. (2000). "Female feral fowl eject sperm of subdominant males". Nature. 405: 787–789. Bibcode:2000Natur.405..787P. doi:10.1038/35015558.
34. Clutton-Brock, T.H. (1991). The Evolution of Parental Care. Princeton NJ: Princeton University Press.
35. Wcislo, W. T.; Wille, A.; Orozco, E. (1993). "Nesting biology of tropical solitary and social sweat bees, Lasioglossum (Dialictus) figueresi Wcislo and L. (D.) aeneiventre (Friese) (Hymenoptera: Halictidae)". Insectes Sociaux. 40: 21–40. doi:10.1007/BF01338830.
36. Daly, M. (1979). "Why Don't Male Mammals Lactate?". Journal of Theoretical Biology. 78 (3): 325–345. doi:10.1016/0022-5193(79)90334-5. PMID 513786.

37. Gross, M.R.; R.C Sargent (1985). "The evolution of male and female parentental care in fishes". American Zoologist. 25: 807–822. doi:10.1093/icb/25.3.807.
38. Foote, Chris J; Brown, Gayle S; Hawryshyn, Craig W (1 January 2004). "Female colour and male choice in sockeye salmon: implications for the phenotypic convergence of anadromous and nonanadromous morphs". Animal Behaviour. 67 (1): 69–83. doi:10.1016/j.anbehav.2003.02.004.
39. Svensson; Magnhagen, C. (Jul 1998). "Parental behavior in relation to the occurrence of sneaking in the common goby". Animal Behaviour. 56 (1): 175–179. doi:10.1006/anbe.1998.0769. PMID 9710475.
40. Clutton-Brock, T. H. (1991). "The evolution of parental care". Princeton University Press.
41. Sturmbauer, Christian; Corinna Fuchs; Georg Harb; Elisabeth Damm; Nina Duftner; Michaela Maderbacher; Martin Koch; Stephan Koblmüller (2008). "Abundance, Distribution, and Territory Areas of Rock-dwelling Lake Tanganyika Cichlid Fish Species". Hydrobiologia. 615 (1): 57–68. doi:10.1007/s10750-008-9557-z. Retrieved 30 September 2013.
42. Johnstone, R.A.; Hinde, C.A. (2006). "Negotiation over offspring care--how should parents respond to each other's efforts?". Behavioral Ecology. 17 (5): 818–827. doi:10.1093/beheco/arl009.
43. Amundsen, T.; Slagsvold, T. (1996). "Lack's Brood Reduction Hypothesis and Avian Hatching Asynchrony: What's Next?". Oikos. 76 (3): 613–620. doi:10.2307/3546359. JSTOR 3546359.
44. Pijanowski, B. C. (1992). "A Revision of Lack's Brood Reduction Hypothesis". The American Naturalist. 139 (6): 1270–1292. doi:10.1086/285386.
45. Trivers, Robert L.; Willard, Dan E. (1976). "Natural selection of parental ability to vary the sex ratio of offspring". Science. 179 (191): 90–92. Bibcode:1973Sci...179...90T. doi:10.1126/science.179.4068.90. PMID 4682135.
46. Bourke, A.F.G. & F.L.W. Ratnieks (2001). "Kin-selected conflict in the bumble-bee Bombus terrestris (Hymenoptera: Apidae)". Proceedings of the Royal Society of London B. 268: 347–355. doi:10.1098/rspb.2000.1381.
47. Haig, D.; Graham, C. (1991). "Genomic imprinting and the strange case of the insulin-like growth factor-II receptor". Cell. 64: 1045–1046. doi:10.1016/0092-8674(91)90256-x.
48. Kilner, R. M. (2001). "A Growth Cost of Begging in Captive Canary Chicks". Proceedings of the National Academy of Sciences of the United States of America. 98 (20): 11394–11398. Bibcode:2001PNAS...9811394K. doi:10.1073/pnas.191221798.
49. Kolliker, M.; Brinkhof, M.; Heeb, P.; Fitze, P.; Richner, H. (2000). "The Quantitative Genetic Basis of Offspring Solicitation and Parental Response in a Passerine Bird with Parental Care". Proceedings of the Royal Society B: Biological Sciences. 267 (1457): 2127–2132. doi:10.1098/rspb.2000.1259.
50. Trillmitch, F.; Wolf, J.B.W. (2008). "Parent-offspring and sibling conflict in

Galapagos fur seals and sea lions". Behavioral Ecology and Sociobiology. 62 (3): 363–375. doi:10.1007/s00265-007-0423-1.

51. Drummond, H.; Chavelas, C.G. (1989). "Food shortage influences sibling sggression in the Blue-footed Booby". Animal Behaviour. 37: 806–819. doi:10.1016/0003-3472(89)90065-1.

52. Briskie, James V.; Naugler, Christopher T.; Leech, Susan M. (1994). "Begging intensity of nestling birds varies with sibling relatedness". Proceedings of the Royal Society B: Biological Sciences. 258 (1351): 73–78. Bibcode:1994RSPSB.258...73B. doi:10.1098/rspb.1994.0144.

53. Spottiswoode, C. N.; Stevens, M. (2010). "Visual modelling shows that avian host parents use multiple visual cues in rejecting parasitic eggs". Proceedings of the National Academy of Sciences of the United States of America. 107 (19): 8672–8676. Bibcode:2010PNAS..107.8672S. doi:10.1073/pnas.0910486107.

54. Kliner, R.M.; Madden, Joah R.; Hauber, Mark E. (2004). "Brood parasitic cowbird nestlings use host young to procure resources". Science. 305 (5685): 877–879. Bibcode:2004Sci...305..877K. doi:10.1126/science.1098487.

55. Thomas, J.A.; Settele, Josef (2004). "Butterfly mimics of ants". Nature. 432 (7015): 283–284. Bibcode:2004Natur.432..283T. doi:10.1038/432283a.

56. Davies, N.B. (2011). "Cuckoo adaptations: trickery and tuning". Journal of Zoology. 281: 1–14. doi:10.1111/j.1469-7998.2011.00810.x.

57. Tanaka, K. D.; Ueda, K. (2005). "Horsfield's hawk-cuckoo nestlings simulate multiple gapes for begging". Science. 308 (5722): 653. doi:10.1126/science.1109957.

58. Akino, T; J. J. Knapp; J. A. Thomas; G. W. Elmes (1999). "Chemical mimicry and host specificity in the butterfly Maculinea rebeli, a social parasite of Myrmica ant colonies". Proceedings of the Royal Society B. 266 (1427): 1419–1426. doi:10.1098/rspb.1999.0796. Retrieved 28 September 2013.

59. Thomas, Jeremy; Karsten Schönrogge; Simona Bonelli; Francesca Barbero; Emilio Balletto (2010). "Corruption of ant acoustical signals by mimetic social parasites". Communicative and Integrative Biology. 3 (2): 169–171. doi:10.4161/cib.3.2.10603. PMC 2889977 . PMID 20585513.

60. Dapporto L., Cervo R., Sledge M. F., Turillazzi S. (2004) "Rank integration in dominance hierarchies of host colonies by the paper wasp social parasite Polistes sulcifer (Hymenoptera, Vespidae)". J Insect Physiol 50 :217–223

61. Kreuter, Kirsten; Bunk, Elfi (2011). "How the social parasitic bumblebee Bombus bohemicus sneaks into power of reproduction". Behavioral Ecology and Sociobiology. 66 (3): 475–486. doi:10.1007/s00265-011-1294-z.

62. Cameron, Sydney A.; Ramírez, Santiago (2001-07-01). "Nest Architecture and Nesting Ecology of the Orchid Bee Eulaema meriana (Hymenoptera: Apinae: Euglossini)". Journal of the Kansas Entomological Society. 74 (3): 142–165. JSTOR 25086012.

63. Evans, Howard E. "Burrow sharing and nest transfer in the digger wasp Philanthus

gibbosus (Fabricius)". Animal Behaviour. 21 (2): 302–308. doi:10.1016/s0003-3472(73)80071-5.

64. Reed, H. C.; Akre, R. D.; Garnett, W. B. (1979). "A North American Host of the Yellowjacket Social Parasite Vespula austriaca (Panzer) (Hymenoptera: Vespidae)". Entomological News. 90 (2): 110–113.
65. Gjershaug, Jan Ove (2009). "The social parasite bumblebee Bombus hyperboreus Schönherr, 1809 usurp nest of Bombus balteatus Dahlbom, 1832 (Hymenoptera, Apidae) in Norway". Norwegian Journal of Entomology. 56 (1): 28–31.
66. Emlen, S. T.; Oring, L. W. (1977). "Ecology, sexual selection, and the evolution of mating systems". Science. 197: 214–223. Bibcode:1977Sci...197..215E. doi:10.1126/science.327542. PMID 327542.
67. Davies, N.B., Krebs, J.R. and West., S.A., (2012). An Introduction to Behavioural Ecology. 4th ed. John Wiley & Sons, pp. 254-263
68. Warner, R. R. (1987). "Female choice of sites versus mates in a coral reef fish Thalassoma bifasciatum". Animal Behaviour. 35: 1470–1478. doi:10.1016/s0003-3472(87)80019-2.
69. Ims, R.A. (1987). "Responses in spatial organization and behavior to manipulations of the food resource in the vole Clethrionomys rufocanus". Journal of Animal Ecology. 56: 585–596. doi:10.2307/5070.
70. Kimsey, Lynn Siri (1980). "The behaviour of male orchid bees (Apidae, Hymenoptera, Insecta) and the question of leks". Animal Behaviour. 28 (4): 996–1004. doi:10.1016/s0003-3472(80)80088-1.
71. Bradbury, J. E. and Gibson, R. M. (1983) Leks and mate choice. In: Mate Choice (ed. P. Bateson). pp.109-138. Cambridge University Press. Cambridge
72. Moore, Sandra D. (1987). "Male-Biased Mortality in the Butterfly Euphydryas editha: a Novel Cost of Mate Acquisition". The American Naturalist. 130 (2): 306–309. doi:10.1086/284711.
73. Lack, D. (1968) Ecological Adaptations for Breeding in Birds. Methuen, London.
74. Davies, N. B., Krebs, J. R and West, S. A., (2012). An Introduction to Behavioral Ecology. West Sussex, UK: Wiley-Blackwell. pp. 266. ISBN 978-1-4051-1416-5.
75. Lightbody, J.P.; Weaatherhead, P.J. (1988). "Female settling patterns and polygyny: tests of a neutral-mate-choice hypothesis". American naturalist. 132: 20–33. doi:10.1086/284835.
76. Verner, J.; Wilson, M.F. (1966). "The influence of habitats on mating systems of North American passerine birds". Ecology. 47: 143–147. doi:10.2307/1935753.
77. Gowaty, P.A. (1981). "An extension of the Orians-Verner-Willson model to account for mating systems besides polygyny". American Naturalist. 118: 851–859. doi:10.1086/283875.
78. Faaborg, J.; Parker, P.G.; DeLay, L.; et al. (1995). "Confirmation of cooperative polyandry in the Galapagos hawk Buteo galapagoensis". Behavioral Ecology and Sociobiology. 36: 83–90. doi:10.1007/bf00170712.

79. Beissinger, S. R.; Snyder, N. F. R. (1987). "Mate desertion in the snail kite". Animal Behaviour. 35: 477–487. doi:10.1016/s0003-3472(87)80273-7.

80. (Reynolds)

81. Butchart, S. H. M.; Seddon, N.; Ekstrom, J. M. M. (1999b). "Yelling for sex: harem males compete for female access in bronze-winged jacanas". Animal Behaviour. 57: 637–646. doi:10.1006/anbe.1998.0985.

82. Davies, Nicholas B.; Krebs, John R.; West, Stuart A. (2012). An Introduction to Behavioral Ecology. West Sussex, UK: Wiley-Blackwell. pp. 307–333. ISBN 978-1-4051-1416-5.

83. Bergstrom, Theodore (Spring 2002). "Evolution of Social Behavior: Individual and Group Selection". The Journal of Economic Perspectives. 16 (2): 67–88. doi:10.1257/0895330027265. JSTOR 2696497.

84. Smith, J. M. (1964). "Group Selection and Kin Selection". Nature. 201 (4924): 1145–1147. Bibcode:1964Natur.201.1145S. doi:10.1038/2011145a0.

85. Fisher, R. A. (1930). The Genetical Theory of Natural Selection. Oxford: Clarendon Press.

86. Haldane, J.B.S. (1932). The Causes of Evolution. London: Longmans, Green & Co.

87. Haldane, J. B. S. (1955). "Population Genetics". New Biology. 18: 34–51.

88. Hamilton, W. D. (1963). "The evolution of altruistic behavior". American Naturalist. 97 (896): 354–356. doi:10.1086/497114.

89. Hamilton, W. D. (1964). "The Genetical Evolution of Social Behavior". Journal of Theoretical Biology. 7 (1): 1–16. doi:10.1016/0022-5193(64)90038-4. PMID 5875341.

90. West, S.A.; Griffin, A.S.; Gardner, A. (2007b). "Social semantics: altruism, cooperation, mutualism, strong reciprocity and group selection". Journal of Evolutionary Biology. 20 (2): 415–432. doi:10.1111/j.1420-9101.2006.01258.x. PMID 17305808.

91. Mehdiabadi, N. J., C. N. Jack, T. T. Farnham et al. (2006). "Kin preference in a social microbe". Nature. 442 (7105): 881–882. Bibcode:2006Natur.442..881M. doi:10.1038/442881a. PMID 16929288.

92. Benabentos, R., S. Hirose, R. Sucgang et al. (2009). "Polymorphic members of the lag-gene family mediate kin discrimination inDictyostelium". Current Biology. 19 (7): 567–572. doi:10.1016/j.cub.2009.02.037. PMC 2694408 . PMID 19285397.

93. Greenberg, Les (1988-07-01). "Kin recognition in the sweat bee, Lasioglossum zephyrum". Behavior Genetics. 18 (4): 425–438. doi:10.1007/BF01065512. ISSN 0001-8244.

94. Nunes, T. M.; Turatti, I. C. C.; Mateus, S.; Nascimento, F. S.; Lopes, N. P.; Zucchi, R. (2009). "Cuticular Hydrocarbons in the Stingless Bee Schwarziana quadripunctata (Hymenoptera, Apidae, Meliponini): Differences between Colonies, Castes and Age" (PDF). Genetics and Molecular Research. 8 (2): 589–595.

doi:10.4238/vol8-2kerr012. PMID 19551647.

95. Buchwald, Robert; Breed, Michael D. (2005). "Nestmate recognition cues in a stingless bee, Trigona fulviventris". Animal Behaviour. Elsevier. 70 (6): 1331–1337. doi:10.1016/j.anbehav.2005.03.017.
96. Seidelmann, Karsten (2006-09-01). "Open-cell parasitism shapes maternal investment patterns in the Red Mason bee Osmia rufa". Behavioral Ecology. 17 (5): 839–848. doi:10.1093/beheco/arl017. ISSN 1045-2249.
97. Davies, N. B. & M. de L. Brooke (1988). "Cuckoos versus reed warblers: Adaptations and counteradaptations". Animal Behaviour. 36 (1): 262–284. doi:10.1016/S0003-3472(88)80269-0.
98. Holmes, W.G & P.W. Sherman (1982). "The ontogeny of kin recognition in two species of ground squirrels". American Zoologist. 22: 491–517. doi:10.1093/icb/22.3.491.
99. Gloag, R.; et al. (2008). "Nest defence in a stingless bee: What causes fighting swarms in Trigona carbonaria (Hymenoptera, Meliponini)?". Insectes Sociaux Insect. Soc. 55 (4): 387–391. doi:10.1007/s00040-008-1018-1.
100. Wilkinson, G.S. (1984). "Reciprocal food sharing in the vampire bat". Nature. 308 (5955): 181–184. Bibcode:1984Natur.308..181W. doi:10.1038/308181a0.
101. O'Brien, Timothy G. & John G. Robinson (1991). "Allomaternal Care by Female Wedge-Capped Capuchin Monkeys: Effects of Age, Rank and Relatedness". Behaviour. 119: 30–50. doi:10.1163/156853991X00355.
102. Ratnieks, Francis L. W.; Heikki Helanterä (October 2009). "The evolution of extreme altruism and inequality in insect societies". Philosophical Transactions of the Royal Society B. 364 (1553): 3169–3179. doi:10.1098/rstb.2009.0129. Retrieved 28 November 2012.
103. Postgate, J (1998). Nitrogen Fixation, 3rd Edition. Cambridge University Press, Cambridge UK.
104. Dawkins, Richard (1976). The Selfish Gene. Oxford University Press.
105. M.C. Soares; I.M. Côté; S.C. Cardoso & R.Bshary (August 2008). "The cleaning goby mutualism: a system without punishment, partner switching or tactile stimulation". Journal of zoology. 276 (3): 306–312. doi:10.1111/j.1469-7998.2008.00489.x.
106. Foster, Kevin; Tom Wenseleers; Francis L. W. Ranieks (10 September 2001). "Spite: Hamilton's Unproven Theory". Ann. Zool.: 229–238.
107. Strassmann, Joan E.; David C. Queller (May 2007). "Insect societies as divided organisms: The complexities of purpose and cross-purpose". PNAS. 104 (suppl. 1): 8619–8626. Bibcode:2007PNAS..104.8619S. doi:10.1073/pnas.0701285104. PMC 1876438 . PMID 17494750. Retrieved 4 December 2012.
108. Duffy, Emmett J.; Cheryl L. Morrison; Kenneth S. Macdonald (April 2002). "Colony defense and behavioral differentiation in the eusocial shrimp Synalpheus regalis". Behavioral Ecology and Sociobiology. 51 (5): 488–495. doi:10.1007/s00265-002-0455-5. Retrieved 4 December 2012.

109. Foster, W.A. (December 1990). "Experimental evidence for effective and altruistic colony defence against natural predators by soldiers of the gall-forming aphid Pemphigus spyrothecae (Hemiptera: Pemphigidae)". Behavioral Ecology and Sociobiology. 27 (6): 421–430. doi:10.1007/BF00164069. Retrieved 4 December 2012.

110. Jarvis, Jennifer U. M.; Nigel C. Bennett; Andrew C. Spinks (January 1998). "Food availability and foraging by wild colonies of Damaraland mole-rats (Cryptomys damarensis): implications for sociality". Oecologia. 113 (2): 290–298. doi:10.1007/s004420050380. Retrieved 4 December 2012.

111. Bonckaert, W.; Tofilski, A.; Nascimento, F.S.; Billen, J.; Ratnieks, F.L.W.; Wenseleers, T. (2001). "Co-occurrence of three types of egg policing in the Norwegian wasp Dolichovespsula wasp". Behavioral Ecology and Sociobiology. 65: 633–640. doi:10.1007/s00265-010-1064-3.

112. Wenseleers, Tom; Heikki Helanterä; Adam G. Hart; Francis L. W. Ratnieks (May 2004). "Worker reproduction and policing in insect societies: an ESS analysis". Journal of Evolutionary Biology. 17 (5): 1035–1047. doi:10.1111/j.1420-9101.2004.00751.x. PMID 15312076. Retrieved 4 December 2012.

113. Foster, Kevin R. (2001). "Colony kin structure and male production inDolichovespula wasps". Molecular Ecology. 10: 1003–1010. doi:10.1046/j.1365-294X.2001.01228.x. Retrieved 10 October 2014.

114. Vespula vulgaris#Defensive behaviors

115. Andrew F. G. Bourke (1999). "Sex allocation in a facultatively polygynous ant: between-population and between-colony variation"(PDF). Behavioral Ecology. 10 (4): 409–421. doi:10.1093/beheco/10.4.409.

116. Jurgen Heinze; Lipski, Norbert; Schlehmeyer, Kathrin; Hōlldobler, Bert (1994). "Colony structure and reproduction in the ant, Leptothorax acervorum" (PDF). Behavioral Ecology. 6 (4): 359–367. doi:10.1093/beheco/6.4.359.

117. Ratnieks, Francis L.W.; P. Kirk Visscher (December 1989). "Worker policing in the honeybee". Nature. 342 (6251): 796–797. Bibcode:1989Natur.342..796R. doi:10.1038/342796a0. Retrieved 26 November 2012.

118. Gobin, Bruno; J. Billen; C. Peeters (November 1999). "Policing behaviour towards virgin egg layers in a polygynous ponerine ant". Anim. Behav. 58 (5): 1117–1122. doi:10.1006/anbe.1999.1245. PMID 10564615.

119. Boomsma, J.J (21 August 2007). "Kin selection versus sexual selection: why the ends to not meet". Current Biology. 17 (16): R673–R683. doi:10.1016/j.cub.2007.06.033. PMID 17714661.

120. Raina, Ashok K.; Klun, Jerome A. (1984). "Brain factor control of sex pheromone production in the female corn earworm moth". Science. 225 (4661): 531–533. doi:10.1126/science.225.4661.531.

121. O'Donnell, Sean (1997). "Gaster-Flagging during Colony Defense in Neotropical Swarm-Founding Wasps (Hymenoptera: Vespidae, Epiponini)". Journal of the Kansas Entomological Society.

122. Lederhouse, Robert C.; Silvio, G. Codella Jr (1989). "Intersexual Comparison of Mimetic Protection in the Black Swallowtail Butterfly, Papilio polyxenes: Experiments with Captive Blue Jay Predators". Evolution. 43 (2): 410–420. doi:10.2307/2409216.

CHAPTER 7

Primatology

Chapter Contents

- **Introduction**
- **Sub-Disciplines**
- **Japanese Primatology**
- **Women in Primatology**

Primatology is the scientific study of living non-human primates. Primates are our closest relatives among all living things. Monkeys, gibbons, greater apes and humans are all primates. Primatology exists because humans have a unique place within primates. Primatology gives us a greater understanding of biological nature.

INTRODUCTION

Primatology is the scientific study of primates. It is a diverse discipline and researchers can be found in academic departments of anatomy, anthropology, biology, medicine, psychology, veterinary sciences and zoology, as well as in animal sanctuaries, biomedical research facilities, museums and zoos. Primatologists study both living and extinct primates in their natural habitats and in laboratories by conducting field studies and experiments in order to understand aspects of their evolution and behaviour. Primatologists often divide primates into three groups for study: dominant females, females and young, and peripheral males.

Olive baboon

SUB-DISCIPLINES

As a science, primatology has many different sub-disciplines which vary in terms of theoretical and methodological approaches to the subject used in researching extant primates and their extinct ancestors.

There are two main centers of primatology, Western primatology and Japanese primatology. These two divergent disciplines stem from their unique cultural backgrounds and philosophies that went into their founding. Although, fundamentally, both Western and Japanese primatology share many of the same principles, the areas of their focus in primate research and their methods of obtaining data differ widely.

Western Primatology

Origins

Western primatology stems primarily from research by North American and European scientists. Early primate study focused primarily in medical research, but some scientists also conducted "civilizing" experiments on chimpanzees in order to gauge both primate intelligence and the limits of their brainpower...

Theory

The study of primatology looks at the biological and psychological aspects of non-human primates. The focus is on studying the common links between humans and primates. It is believed that by understanding our closest animal relatives, we might better understand the nature shared with our ancestors.

Methods

Primatology is a science. The general belief is that the scientific observation of nature must be either extremely limited, or completely controlled. Either way, the observers must be neutral to their subjects. This allows for data to be unbiased and for the subjects to be uninfluenced by human interference.

There are three methodological approaches in primatology: field study, the more realistic approach, laboratory study, the more controlled approach, and semi-free ranging, where primate habitat and wild social structure is replicated in a captive setting.

Field is done in natural environments, in which scientific observers watch primates in their natural habitat.

Laboratory study is done in controlled lab settings. In lab settings, scientists are able to perform controlled experimentation on the learning capabilities and behavioral patterns of the animals.

In semi-free ranging studies, scientists are able to watch how primates might act in the wild but have easier access to them, and the ability to control their environments. Such facilities include the Living Links Center at the Yerkes National Primate Research Center in Georgia and the Elgin Center at Lion Country Safari in Florida.

All types of primate study in the Western methodology are meant to be neutral. Although there are certain Western primatologists who do more subjective research, the emphasis in this discipline is on the objective.

Early field primatology tended to focus on individual researchers. Researchers such as Dian Fossey and Jane Goodall are examples of this. Long-term sites of research tend to be best associated with their founders, and this led to some tension between younger primatologists and the veterans in the field.

Notable Western Primatologists

- Nina Amstrup
- Jeanne Altmann
- Jean Baulu
- Irwin Bernstein
- Michelle Bezanson
- Sarah Blaffer Hrdy
- Christophe Boesch
- Sue Boinski
- Geoffrey Bourne
- Josep Call
- C. R. Carpenter

- Colin Chapman
- Dorothy Cheney
- Marina Cords
- Thomas Defler
- Frans de Waal
- Alan F. Dixson
- Joseph Erwin
- Alejandro Estrada
- Linda Fedigan
- Dian Fossey
- Dorothy Fragaszy
- Agustin Fuentes
- Birutė Galdikas
- Paul Garber
- Thomas R. Gillespie
- Jane Goodall
- Colin Groves
- Alexander Harcourt
- Harry Harlow
- Philip Hershkovitz
- William C. Osman Hill
- Gottfried Hohmann
- Lynne Isbell
- Alison Jolly
- Hans Kummer
- Boris Lapin
- Nadezhda Ladygina-Kohts
- Louis Leakey
- Donald Lindburg
- Robert D. Martin
- William Mason
- Bill McGrew
- Emil Wolfgang Menzel, Jr.
- Gary D. Mitchell
- John Mitani
- Russell Mittermeier

- John R. Napier
- Prudence H. Napier
- Nicholas Newton-Fisher
- William C. Osman Hill
- Matthew Richardson
- Shawn Ridgeway
- Anne E. Russon
- Anthony Rylands
- Jordi Sabater Pi
- Gene P. Sackett
- Robert Sapolsky
- Carel van Schaik
- Adolph H. Schultz
- Robert M. Seyfarth
- Joan Silk
- Meredith Small
- Barbara Smuts
- Charles Southwick
- Craig Stanford
- Karen Strier
- Tom Struhsaker
- Robert W. Sussman
- Michael Tomasello
- Omar Wasow
- Sherwood Washburn
- David P. Watts
- Barbara J. Welker
- Richard Wrangham
- Gabriel Zunino
- Susan Emily Perry
- Carlos A. Peres

JAPANESE PRIMATOLOGY

Origins

The discipline of Japanese primatology was developed out of animal ecology. It is mainly credited to Kinji Imanishi and Junichiro Itani. Imanishi was an animal ecologist who began studying wild horses before focusing more on primate ecology. He helped found the Primate Research Group in 1950. Junichiro was a renowned anthropologist and a professor at Kyoto University. He is a co-founder of the Primate Research Institute and the Centre for African Area Studies.

Theory

The Japanese discipline of primatology tends to be more interested in the social aspects of primates. Social evolution and anthropology are of primary interest to them. The Japanese theory believes that studying primates will give us insight into the duality of human nature: individual self vs. social self.

The traditional and cultural aspects of Japanese science lend themselves to an "older sibling" mentality. It is believed that animals should be treated with respect, but also a firm authority. This is not to say that the Japanese study of primatology is cruel – far from it – just that it does not feel that their subjects should be given reverential treatment.

One particular Japanese primatologist, Kawai Masao, introduced the concept of *kyokan*. This was the theory that the only way to attain reliable scientific knowledge was to attain a mutual relation, personal attachment and shared life with the animal subjects. Though Kawai is the only Japanese primatologist associated with the use of this term, the underlying principle is part of the foundation of Japanese primate research.

Methods

Japanese primatology is a carefully disciplined subjective science. It is believed that the best data comes through identification with your subject. Neutrality is eschewed in favour of a more casual atmosphere, where researcher and subject can mingle more freely. Domestication of nature is not only desirable, but necessary for study.

Japanese primatologists are renowned for their ability to recognise animals by sight, and indeed most primates in a research group are usually named and numbered. Comprehensive data on every single subject in a group is uniquely Japanese trait of primate research. Each member of the primate community has a part to play, and the Japanese researchers are interested in this complex interaction.

For Japanese researchers in primatology, the findings of the team are emphasised over the individual. The study of primates is a group effort, and the group will get the credit for it. A team of researchers may observe a group of primates for several years in order to gather very detailed demographic and social histories.

Notable Japanese Primatologists

- Imanishi, Kinji
- Junichiro, Itani
- Masao, Kawai
- Tetsuro Matsuzawa

- Toshisada Nishida
- Hiraiwa-Hasegawa, Mariko
- Jensen, Erik
- Satsue Mito

Primatology in Sociobiology

Where sociobiology attempts to understand the actions of all animal species within the context of advantageous and disadvantageous behaviors, primatology takes an exclusive look at the order Primates, which includes *Homo sapiens*. The interface between primatology and sociobiology examines in detail the evolution of primate behavioral processes, and what studying our closest living primate relatives can tell about our own minds. As the American anthropologist Earnest Albert Hooton used to say: "Primas sum: primatum nil a me alienum puto" ("I am a primate; nothing about primates is outside of my bailiwick"). The meeting point of these two disciplines has become a nexus of discussion on key issues concerning the evolution of sociality, the development and purpose of language and deceit, and the development and propagation of culture.

Additionally, this interface is of particular interest to the science watchers in science and technology studies, who examine the social conditions which incite, mould, and eventually react to scientific discoveries and knowledge. The STS approach to primatology and sociobiology stretches beyond studying the apes, into the realm of observing the people studying the apes.

Taxonomic Basis

Before Darwin, and before molecular biology, the father of modern taxonomy, Carl Linnaeus, organized natural objects into kinds, that we now know reflect their evolutionary relatedness. He sorted these kinds by morphology, the shape of the object. Animals such as monkeys, chimpanzees and orangutans resemble humans closely, so Linnaeus placed *Homo sapiens* together with other similar-looking organisms into the taxonomic order *Primates*. Modern molecular biology reinforced humanity's place within the Primate order. Humans and simians share the vast majority of their DNA, with chimpanzees sharing between 97-99% genetic identity with humans.

From Grooming to Speaking

Although social grooming is observed in many animal species, the grooming activities undertaken by primates are not strictly for the elimination of parasites. In primates, grooming is a social activity that strengthens relationships. The amount of grooming taking place between members of a troop is a potent indicator of alliance formation or troop solidarity. Robin Dunbar suggests a link between primate grooming and the development of human language. The size of the neocortex in a primate's brain correlates directly to the number of individuals it can keep track of socially, be it a troop of chimps or a tribe of humans.

This number is referred to as the monkeysphere. If a population exceeds the size outlined by its cognitive limitations, the group undergoes a schism. Set into an evolutionary context, the Dunbar number shows a drive for the development of a method of bonding that is less labor-intensive than grooming: language. As the monkeysphere grows, the amount of time that would need to be spent grooming troopmates soon becomes unmanageable. Furthermore, it is only possible to bond with one troopmate at a time while grooming. The evolution of vocal communication solves both the time constraint and the one-on-one problem, but at a price.

Language allows for bonding with multiple people at the same time at a distance, but the bonding produced by language is less intense. This view of language evolution covers the general biological trends needed for language development, but it takes another hypothesis to uncover the evolution of the cognitive processes necessary for language.

Modularity of the Primate Mind

Noam Chomsky's concept of innate language addresses the existence of universal grammar, which suggests a special kind of "device" all humans are born with whose sole purpose is language. Fodor's modular mind hypothesis expands on this concept, suggesting the existence of preprogrammed modules for dealing with many, or all aspects of cognition. Although these modules do not need to be physically distinct, they must be functionally distinct. Orangutans are currently being taught language at the Smithsonian National Zoo using a computer system developed by primatologist Dr. Francine Neago in conjunction with IBM.

The massive modularity theory thesis posits that there is a huge number of tremendously interlinked but specialized modules running programs called Darwinian algorithms, or DA. DA can be selected for just as a gene can, eventually improving cognition. The contrary theory, of generalist mind, suggests that the brain is just a big computer that runs one program, the mind. If the mind is a general computer, for instance, the ability to use reasoning should be identical regardless of the context. This is not what is observed. When faced with abstract numbers and letters with no "real world" significance, respondents of the Wason card test generally do very poorly. However, when exposed to a test with an identical rule set but socially relevant content, respondents score markedly higher. The difference is especially pronounced when the content is about reward and payment. This test strongly suggests that human logic is based on a module originally developed in a social environment to root out cheaters, and that either the module is at a huge disadvantage where abstract thinking is involved, or that other less effective modules are used when faced with abstract logic.

Further evidence supporting the modular mind has steadily emerged with some startling revelations concerning primates. A very recent study indicated that human babies and grown monkeys approach and process numbers in a similar fashion, suggesting an evolved set of DA for mathematics (Jordan). The conceptualization of both human infants and primate adults is cross-sensory, meaning that they can add 15 red dots to 20 beeps and approximate the answer to be 35 grey squares. As more evidence of basic cognitive modules are uncovered, they will undoubtedly form a more solid foundation upon which the more complex behaviors can be understood.

In contradiction to this, neuroscientist Jaak Panksepp has argued that the mind is not a computer nor is it massively modular. He states that no evidence of massive modularity or the brain as a digital computer has been gained through actual neuroscience, as opposed to psychological studies. He criticises psychologists who use the massive modularity thesis for not integrating neuroscience into their understanding.

The Primate Theory of Mind

Primate behavior, like human behavior, is highly social and ripe with the intrigue of kingmaking, powerplays, deception, cuckoldry, and apology. In order to understand the staggeringly complex nature of primate interactions, we look to theory of mind. Theory of mind asks whether or not an individual recognizes and can keep track of information asymmetry amongst individuals in the group, and whether or not they can attribute folk psychological states to their peers. If some primates can tell what others know and want and act accordingly, they can gain advantage and status.

Recently, chimpanzee theory of mind has been advanced by Felix Warneken of the Max Planck Institute. His studies have shown that chimpanzees can recognize whether a researcher desires a dropped object, and act accordingly by picking it up. Even more compelling is the observation that chimps will only act if the object is dropped in an accidental-looking manner: if the researcher drops the object in a way that appears intentional, the chimp will ignore the object.

In a related experiment, groups of chimps were given rope-pulling problems they could not solve individually. Warneken's subjects rapidly figured out which individual in the group was the best rope puller and assigned it the bulk of the task. This research is highly indicative of the ability of chimps to detect the folk psychological state of "desire", as well as the ability to recognize that other individuals are better at certain tasks than they are.

However primates do not always fare so well in situations requiring theory of mind. In one experiment pairs of chimpanzees who had been close grooming partners were offered two levers. Pressing one lever would bring them food and another would bring their grooming partner food. Pressing the lever to clearly give their grooming partner much-wanted food would not take away from how much food they themselves got. For some reason, the chimps were unwilling to depress the lever that would give their long-time chums food. It is plausible but unlikely that the chimps figured there was finite food and it would eventually decrease their own food reward. The experiments are open to such interpretations making it hard to establish anything for certain.

One phenomenon which would indicate a possible fragility of theory of mind in primates occurs when a baboon gets lost. Under such circumstances, the lost baboon generally makes "call barks" to announce that it is lost. Previous to the 1990s it was thought that these call barks would then be returned by the other baboons, similar to the case is in vervet monkeys. However, when researchers studied this formally in the past few years they found something surprising: Only the baboons who were lost would ever give call barks. Even if an infant was wailing in agony just a few hundred meters away, its mother who would clearly recognise its voice and would be frantic about his safety (or alternatively run towards her infant depending on her own perceived safety), would often simply stare in his direction visibly agitated. If the anguishing baboon mother made any type of call at all, the infant would instantly recognise her and run to her position. This type of logic appears to be lost on the baboon, suggesting a serious gap in theory of mind of this otherwise seemingly very intelligent primate species. However, it is also possible that baboons do not return call barks for ecological reasons, for example because returning the call bark might call attention to the lost baboon, putting it at greater risk from predators.

Criticisms

Scientific studies concerning primate and human behavior have been subject to the same set of political and social complications, or biases, as every other scientific discipline. The borderline and multidisciplinary nature of primatology and sociobiology make them ripe fields of study because they are amalgams of objective and subjective sciences. Current scientific practice, especially in the hard sciences, requires a total dissociation of personal experience from the finished scientific product (Bauchspies 8). This is a strategy that is incompatible with observational field studies, and weakens them in the eyes of hard science. As mentioned above, the Western school of primatology tries to minimize subjectivity, while the Japanese school of primatology tends to embrace the closeness inherent in studying nature.

Social critics of science, some operating from within the field, are critical of primatology and sociobiology. Claims are made that researchers bring pre-existing opinions on issues concerning human sociality to their studies, and then seek evidence that agrees with their worldview or otherwise furthers a sociopolitical agenda. In particular, the use of primatological studies to assert gender roles, and to both promote and subvert feminism has been a point of contention.

An example of this is Zuckerman's 1932 study of captive hamadryas baboons, as critiqued in Sturm and Fedigan's Changing Views on Primate Societies. Zuckerman observed male baboons kill each other in a captive environment. Whether intended or not, the study served to reinforce images of the male as the sole competitor in an often violent race to secure dominance and access to a harem of females. Despite unrealistic overcrowding and unnatural male-to-female ratios, Zuckerman's paper was viewed as good science at the time. These ideas were used to justify the status quo of human male dominance, and these interpretations were widely accepted and assumed to be the basis of a primate-wide template for behavior, including that of humans. Incidentally, the hamadryas baboon females are among the most submissive and most gender-unequal of all primates, although primates and humans share a tremendous variation in troop structure (Hrdy 101, Stone).

Several research papers on primate cognition were retracted in 2010. Their lead author, primatologist Marc Hauser, was dismissed from Harvard University after an internal investigation found evidence of scientific misconduct in his laboratory. Data supporting the authors' conclusion that cottontop tamarin monkeys displayed pattern-learning behavior similar to human infants reportedly could not be located after a three-year investigation.

WOMEN IN PRIMATOLOGY

Women receive the majority of Ph.D's in primatology. Londa Schiebinger, writing in 2001, estimated that women made up 80 percent of graduate students pursuing Ph.D's in primatology, up from 50 percent in the 1970s. Because of the high number of women, Schiebinger has even asserted that "Primatology is widely celebrated as a feminist science".

Changing Stereotypes

With attention to Darwin's perception about sexual selection, it was perceived that sexual selection acted differently on females and males. Early research emphasized male-male competition for females. It is widely believed that males tend to woo females, and that females were passive. For years this was the dominant interpretation, emphasizing competition among dominant males who controlled territorial boundaries and maintained order among lesser males. Females on the other hand were described as "dedicated mothers to small infants and sexually available to males in order of the males' dominance rank". Female-female competition was ignored. Schiebinger proposed that the failure to acknowledge female-female competitions could "skew notions of sexual selection" to "ignore interactions between males and females that go beyond the strict interpretation of sex as for reproduction only." In the 1960s primatologists started looking at what females did, slowly changing the stereotype of the passive female. We now know that females are active participants, and even leaders, within their groups. For instance, Rowell found that female baboons determine the route for daily foraging (primary ref needed). Similarly, Shirley Strum found that male investment in special relationships with females had greater productive payoff in comparison to a male's rank in a dominance hierarchy (primary ref needed). This emerging "female point of view" resulted in a reanalysis of how aggression, reproductive access, and dominance affect primate societies.

Schiebinger has also accused sociobiologists of producing the "corporate primate", described as "female baboons with briefcases, strategically competitive and aggressive." This contrasts with the notion that only men are competitive and aggressive. Observations have repeatedly demonstrated that female apes and monkeys also form stable dominance hierarchies and alliances with their male counterparts. Females display aggression, exercise sexual choice, and compete for resources, mates and territory, like their male counterparts.

Six Different Features of Feminist Science that Characterize Contemporary Primatology (Fedigan).

1. Reflexivity: sensitivity to context and cultural bias in scientific work.
2. "The female point of view"
3. Respect for nature and an ethic cooperation with nature
4. Move away from reductionism
5. Promote humanitarian values rather than national interests
6. Diverse community, accessible and egalitarian

Schiebinger suggests that only two out of the six features characteristic of feminism. One of them is the discussion of the politics of participation and the attention placed on females as subjects of research.

The Evolution of Primatology

In 1970 Jeanne Altmann drew attention to representative sampling methods in which all individuals, not just the dominant and the powerful, were observed for equal periods of time. Prior to 1970, primatologists used "opportunistic sampling," which only recorded what caught their attention. Sarah Hrdy, a self-identified feminist, was among the first to apply what became known as sociobiological theory to primates. In her studies, she focuses on the need for females to win from males parental care for their offspring. Linda Fedigan views herself as a reporter or translator, working at the intersection between gender studies of science and the mainstream study of primatology. While some influential women challenged fundamental paradigms, Schiebinger suggests that science is constituted by numerous factors varying from gender roles and domestic issues that surround race and class to economic relations between researchers from Developed World countries and the Developing World countries in which most nonhuman primates reside.

Academic Resources

Societies

- American Society of Primatologists
- European Federation for Primatology
- International Primatological Society

Journals

- American Journal of Primatology
- Folia Primatologica
- International Journal of Primatology
- Journal of Medical Primatology
- Journal of Human Evolution
- Primates

REFERENCES

1. "What is Primatology?". Primate Info Net. Retrieved 9 June 2011.
2. "What is a Primatologist?". Primate Info Net. Retrieved 9 June 2011.
3. Schiebinger, Londa (2001). Has Feminism Changed Science?. First Harvard University Press. p. 129. ISBN 0-674-00544-9.

4. http://www.flyfishingdevon.co.uk/salmon/year3/psy364criticisms-evolutionary-psychology/panksepp_seven_sins.pdf
5. Johnson, Carolyn. “Author on Leave After Harvard Inquiry”, The Boston Globe, 10 August 2010. Retrieved 29 August 2010.
6. Schiebinger, Londa (2001). Has Feminism Changed Science?. First Harvard University Press.
7. Schiebinger, Londa (2001). Has Feminism Changed Science. First Harvard University Press.
8. John Wiley & Sons. “American Journal of Primatology”. Retrieved 2 February 2009.
9. Springer. “International Journal of Primatology”. Retrieved 2 February 2009.
10. John Wiley & Sons. “Journal of Medical Primatology”. Retrieved 2 February 2009.
11. Elsevier. “Journal of Human Evolution”. Retrieved 2 February 2009.

CHAPTER 8

Cetalogy

Chapter Contents

- **Introduction**
- **Whale**
- **Dolphin**

Whales were hunted mercilessly in the 18th and 19th centuries for meat and the oil contained in their blubber. Among its uses, whale oil was used for lighting lamps. Whaling ships like the Pequod depicted in Melville's Moby Dick sailed the seas in a search for energy.

INTRODUCTION

Cetology (from Greek κτος, kētos, "whale"; and -λογία, -logia) or Whalelore is the branch of marine mammal science that studies the approximately eighty species of whales, dolphins, and porpoise in the scientific order Cetacea. Cetologists, or those who practice cetology, seek to understand and explain cetacean evolution, distribution, morphology, behavior, community dynamics, and other topics.

WHALE

Whale is the common name for a widely distributed and diverse group of fully aquatic placentalmarine mammals. They are an informal grouping within the infraorder Cetacea, usually excluding dolphins and porpoises. Whales, dolphins and porpoises belong to the order Cetartiodactyla with even-toed ungulates and their closest living relatives are the hippopotamuses, having diverged about 40 million years ago. The two parvorders of whales, baleen whales (Mysticeti) and toothed whales (Odontoceti), are thought to have split apart around 34 million years ago. The whales comprise eight extant families: Balaenopteridae (the rorquals), Balaenidae (right whales), Cetotheriidae (the pygmy right whale), Eschrichtiidae (the gray whale), Monodontidae (belugas and narwhals), Physeteridae (the sperm whale), Kogiidae (the dwarf and pygmy sperm whale), and Ziphiidae (the beaked whales).

Whales are creatures of the open ocean; they feed, mate, give birth, suckle and raise their young at sea. So extreme is their adaptation to life underwater that they are unable to survive on land. Whales range in size from the 2.6 metres (8.5 ft) and 135 kilograms (298 lb) dwarf sperm whale to the 29.9 metres (98 ft) and 190 metric tons (210 short tons) blue whale, which is the largest creature that has ever lived. The sperm whale is the largest toothed predator on earth. Several species exhibit sexual dimorphism, in that the females are larger than males. Baleen whales have no teeth; instead they have plates of baleen, a fringe-like structure used to expel water while retaining the krill and plankton which they feed on. They use their throat pleats to expand the mouth to take in huge gulps of water. Balaenids have heads that can make up 40% of their body mass to take in water. Toothed whales, on the other hand, have conical teeth designed for catching fish or squid. Baleen whales have a well developed sense of "smell", whereas toothed whales have well-developed hearing – their hearing, that is adapted for both air and water, is so well developed that some can survive even if they are blind. Some species, such as sperm whales, are well adapted for diving to great depths to catch squid and other favoured prey.

Whales have evolved from land-living mammals. As such they must breathe air regularly, though they can remain submerged for long periods. They have blowholes (modified nostrils) located on top of their heads, through which air is taken in and expelled in the form of vapour. They are warm-blooded, and have a layer of fat, or blubber, under the skin. With streamlined fusiform bodies and two limbs that are modified into flippers, whales can travel at up to 20 knots, though they are not as flexible or agile as seals. Whales produce a great variety of vocalizations, notably the extended songs of the humpback whale. Although whales are widespread, most species prefer the colder waters of the Northern and Southern Hemispheres, and migrate to the equator to give birth. Species such as humpbacks and blue whales are capable of travelling thousands of miles without feeding. Males typically mate with multiple females every year, but females only mate every two to three years. Calves are typically born in the spring and summer months and females bear all the responsibility for raising them. Mothers of some species fast and nurse their young for a relatively long time.

Once relentlessly hunted for their products, whales are now protected by international law. The North Atlantic right whales nearly became extinct in the twentieth century, with a population low of 450, and the North Pacific gray whale population is ranked Critically Endangered by the IUCN. Besides whaling, they also face threats from bycatch and marine pollution. The meat, blubber and baleen of whales have traditionally been used by indigenous peoples of the Arctic. Whales have been depicted in various cultures worldwide, notably by the Inuit and the coastal peoples of Vietnam and Ghana, who sometimes hold whale funerals. Whales occasionally feature in literature and film, as in the great white whale of Herman Melville's *Moby Dick*. Small whales, such as belugas, are sometimes kept in captivity and trained to perform tricks, but breeding success has been poor and the animals often die within a few months of capture. Whale watching has become a form of tourism around the world.

Southern right whale

Etymology and Definitions

The word “whale” comes from the Old English *whæl*, related to the High German *wal*. Related forms are the Old Norse *hvalr* and the Swedish/Danish *hval*. The obsolete “whalefish” has a similar derivation, indicating a time when whales were thought to be fish. Other archaic English forms include *wal, wale, whal, whalle, whaille, wheal*, etc.

The term “whale” is sometimes used interchangeably with dolphins and porpoises, acting as a synonym for Cetacea. Six species of dolphins have the word “whale” in their name, collectively known as blackfish: the killer whale, the melon-headed whale, the pygmy killer whale, the false killer whale, and the two species of pilot whales, all of which are classified under the family Delphinidae (oceanic dolphins). Each species has a different reason for it, for example, the killer whale was named “Ballena asesina” by Spanish sailors, which translates directly to “whale assassin” or “whale killer”, but is more often translated to “killer whale”.

Taxonomy and Evolution

The hippopotamuses are the sister clade to the Cetacea, which contains whales, dolphins and porpoises.

The whales are part of the largely terrestrial mammalian clade Laurasiatheria. Whales do not form a clade or order; the infraorder Cetacea includes dolphins and porpoises, which are not considered whales.

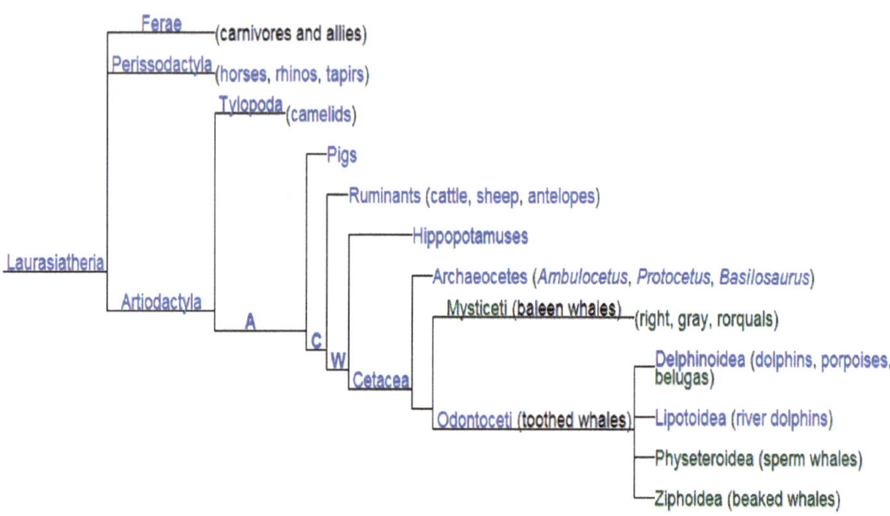

- Whales in green
- **A**=Artiofabula
- **C**=Cetruminantia

- **W**=Whippomorpha

Cetaceans are divided into two parvorders: the largest parvorder, Mysticeti (baleen whales), is characterized by the presence of baleen, a sieve-like structure in the upper jaw made of keratin, which it uses to filter plankton, among others, from the water; Odontocetes (toothed whales) are characterized by bearing sharp teeth for hunting, as opposed to their counterparts' baleen.

Cetaceans and artiodactyls now are classified under the order Cetartiodactyla, often still referred to as Artiodactyla, which includes both whales and hippopotamuses. The hippopotamus and pygmy hippopotamus are the whale's closest terrestrial living relatives.

Mysticetes

Mysticetes are also known as baleen whales. They have a pair of blowholes side-by-side and lack teeth, which renders them incapable of catching larger prey; they instead have baleen plates which is a sieve-like structure in the upper jaw made of keratin, which it uses to filter plankton and other food from the water; this forces them to follow krill or plankton migrations. Some whales, such as the humpback, reside in the polar regions where they feed on a reliable source of schooling fish and krill. These animals rely on their well-developed flippers and tail fin to propel themselves through the water; they swim by moving their fore-flippers and tail fin up and down. Whale ribs loosely articulate with their thoracic vertebrae at the proximal end, but do not form a rigid rib cage. This adaptation allows their chest to compress during deep dives as the pressure increases with depth. Mysticetes consist of four families: rorquals (balaenopterids), cetotheriids, right whales (balaenids), and gray whales (eschrichtiids).

The main difference between each family of mysticete is in their feeding adaptations and subsequent behaviour. Balaenopterids are the rorquals. These animals, along with the cetotheriids, rely on their throat pleats to gulp large amounts of water while feeding. The throat pleats extend from the mouth to the navel and allow the mouth to expand to a large volume for more efficient capture of the small animals they feed on. Balaenopterids consist of two genera and eight species. Balaenids are the right whales. These animals have very large heads, which can make up as much as 40% of their body mass, and much of the head is the mouth. This allows them to take in large amounts of water into their mouths, letting them feed more effectively. Eschrichtiids have one living member: the gray whale. They are bottom feeders, mainly eating crustaceans and benthic invertebrates. They feed by turning on their sides and taking in water mixed with sediment, which is then expelled through the baleen, leaving their prey trapped inside. This is an efficient method of hunting, in which the whale has no major competitors.

Odontocetes

Odontocetes are known as toothed whales; they have teeth and only one blowhole. They rely on their well-developed sonar to find their way in the water. Toothed whales send out ultrasonic clicks using the melon. Sound waves travel through the water. Upon striking an object in the water, the sound waves bounce back at the whale. These vibrations are received through fatty tissues in the jaw, which is then rerouted into the ear-bone and into the brain where the vibrations are interpreted. All toothed whales are opportunistic, meaning they will eat anything they can fit in their throat because they are unable to chew. These animals rely on their well-developed flippers and tail fin to propel themselves through the water; they swim by moving their fore-flippers and tail fin up and down. Whale ribs loosely articulate with their thoracic vertebrae at the proximal end, but they do not form a rigid rib cage. This adaptation allows the chest to compress during deep dives as opposed to resisting the force of water pressure. Excluding dolphins and porpoises, odontocetes consist of four families: belugas and narwhals (monodontids), sperm whales (physeterids), dwarf and pygmy sperm whales (kogiids), and beaked whales (ziphiids). There are six species, sometimes referred to as "blackfish", that are dolphins commonly misconceived as whales: the killer whale, the melon-headed whale, the pygmy killer whale, the false killer whale, and the two species of pilot whales, all of which are classified under the family Delphinidae (oceanic dolphins).

The differences between families of odontocetes include size, feeding adaptations and distribution. Monodontids consist of two species: the beluga and the narwhal. They both reside in the frigid arctic and both have large amounts of blubber. Belugas, being white, hunt in large pods near the surface and around pack ice, their coloration acting as camouflage. Narwhals, being black, hunt in large pods in the aphotic zone, but their underbelly still remains white to remain camouflaged when something is looking directly up or down at them. They have no dorsal fin to prevent collision with pack ice. Physeterids and Kogiids consist of sperm whales. Sperm whales consist the largest and smallest odontocetes, and spend a large portion of their life hunting squid. *P. macrocephalus* spends most of its life in search of squid in the depths; these animals do not require any degree of light at all, in fact, blind sperm whales have been caught in perfect health. The behaviour of Kogiids remains largely unknown, but, due to their small lungs, they are thought to hunt in the photic zone. Ziphiids consist of 22 species of beaked whale. These vary from size, to coloration, to distribution, but they all share a similar hunting style. They use a suction technique, aided by a pair of grooves on the underside of their head, not unlike the throat pleats on the rorquals, to feed.

Evolution

Basilosaurus skeleton

Whales are descendants of land-dwelling mammals of the artiodactyl order (even-toed ungulates). They are related to the *Indohyus*, an extinct chevrotain-like ungulate, from which they split approximately 48 million years ago. Primitive cetaceans, or archaeocetes, first took to the sea approximately 49 million years ago and became fully aquatic 5–10 million years later. What defines an archaeocete is the presence of anatomical features exclusive to cetaceans, alongside other primitive features not found in modern cetaceans, such as visible legs or asymmetrical teeth. Their features became adapted for living in the marine environment. Major anatomical changes included their hearing set-up that channeled vibrations from the jaw to the earbone (*Ambulocetus* 49 mya), a streamlined body and the growth of flukes on the tail (*Protocetus* 43 mya), the migration of the nostrils toward the top of the cranium (blowholes), and the modification of the forelimbs into flippers (*Basilosaurus* 35 mya), and the shrinking and eventual disappearance of the hind limbs (the first odontocetes and mysticetes 34 mya).

Today, the closest living relatives of cetaceans are the hippopotamuses; these share a semi-aquatic ancestor that branched off from other artiodactyls some 60 mya. Around 40 mya, a common ancestor between the two branched off into cetacea and anthracotheres; nearly all anthracotheres became extinct at the end of the Pleistocene 2.5 mya, eventually leaving only one surviving lineage – the hippopotamus.

Whales split into two separate parvorders around 34 mya – the baleen whales (Mysticetes) and the toothed whales (Odontocetes).

Biology

Anatomy

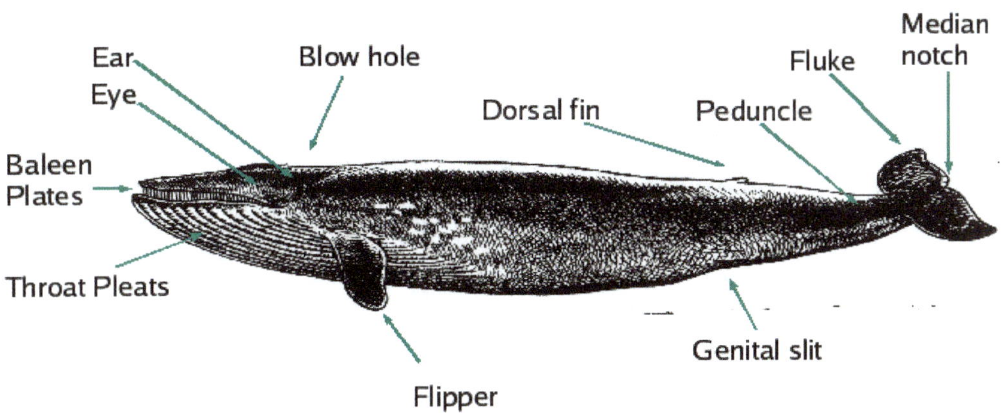

Features of a blue whale

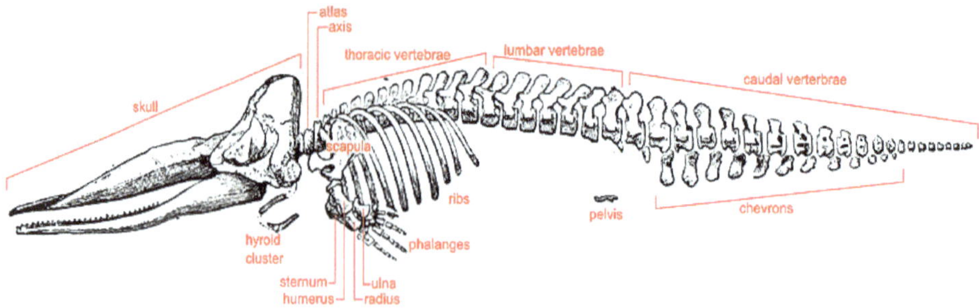

Features of a sperm whale skeleton

Whales have torpedo shaped bodies with non-flexible necks, limbs modified into flippers, non-existent external ear flaps, a large tail fin, and flat heads (with the exception of monodontids and ziphiids). Whale skulls have small eye orbits, long snouts (with the exception of monodontids and ziphiids) and eyes placed on the sides of its head. Whales range in size from the 2.6-metre (8.5 ft) and 135-kilogram (298 lb) dwarf sperm whale to the 34-metre (112 ft) and 190-metric-ton (210-short-ton) blue whale. Overall, they tend to dwarf other cetartiodactyls; the blue whale is the largest creature on earth. Several species have female-biased sexual dimorphism, with the females being larger than the males. One exception is with the sperm whale, which has males larger than the females.

Odontocetes, such as the sperm whale, possess teeth with cementum cells overlying dentine cells. Unlike human teeth, which are composed mostly of enamel on the portion of the tooth outside of the gum, whale teeth have cementum outside the gum. Only in larger whales, where the cementum is worn away on the tip of the tooth, does enamel show. Mysticetes have large whalebone, as opposed to teeth, made of keratin. Mysticetes have two blowholes, whereas Odontocetes contain only one.

Breathing involves expelling stale air from the blowhole, forming an upward, steamy spout, followed by inhaling fresh air into the lungs; a humpback whale's lungs can hold about 5,000 litres of air. Spout shapes differ among species, which facilitates identification.

All whales have a thick layer of blubber. In species that live near the poles, the blubber can be as thick as 11 inches. This blubber can help with buoyancy (which is helpful for a 100-ton whale), protection to some extent as predators would have a hard time getting through a thick layer of fat, and energy for fasting when migrating to the equator; the primary usage for blubber is insulation from the harsh climate. It can constitute as much as 50% of a whales body weight. Calves are born with only a thin layer of blubber, but some species compensate for this with thick lanugos.

Whales have a two- to three-chambered stomach that is similar in structure to terrestrial carnivores. Mysticetes contain a proventriculus as an extension of the oesophagus; this contains stones that grind up food. They also have fundic and pyloric chambers.

Locomotion

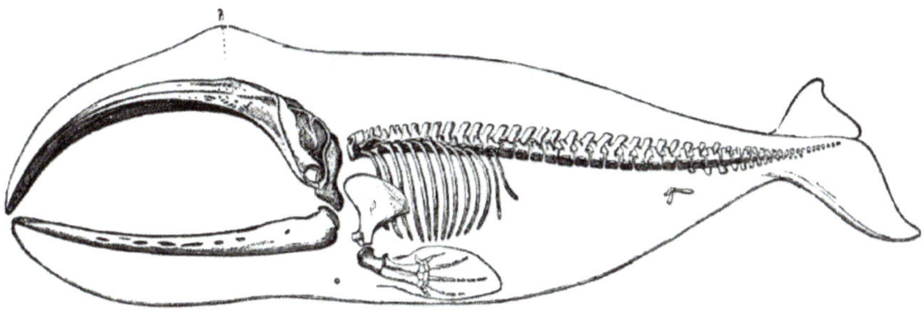

Skeleton of a bowhead whale; notice the hind limb. Richard Lydekker, 1894

Whales have two flippers on the front, and a tail fin. These flippers contain four digits. Although whales do not possess fully developed hind limbs, some, such as the sperm whale and bowhead whale, possess discrete rudimentary appendages, which may contain feet and digits. Whales are fast swimmers in comparison to seals, which typically cruise at 5–15 kn, or 9–28 kilometres per hour (5.6–17.4 mph); the fin whale, in comparison, can travel at speeds up to 47 kilometres per hour (29 mph) and the sperm whale can reach speeds of 35 kilometres per hour (22 mph). The fusing of the neck vertebrae, while increasing stability when swimming at high speeds, decreases flexibility; whales are unable to turn their heads. When swimming, whales rely on their tail fin propel them through the water. Flipper movement is continuous. Whales swim by moving their tail fin and lower body up and down, propelling themselves through vertical movement, while their flippers are mainly used for steering. Some species log out of the water, which may allow then to travel faster. Their skeletal anatomy allows them to be fast swimmers. Most species have a dorsal fin.

Whales are adapted for diving to great depths. In addition to their streamlined bodies, they can slow their heart rate to conserve oxygen; blood is rerouted from tissue tolerant of water pressure to the heart and brain among other organs; haemoglobin and myoglobin store oxygen in body tissue; and they have twice the concentration of myoglobin than haemoglobin. Before going on long dives, many whales exhibit a behaviour known as sounding; they stay close to the surface for a series of short, shallow dives while building their oxygen reserves, and then make a sounding dive.

Senses

Biosonar by cetaceans

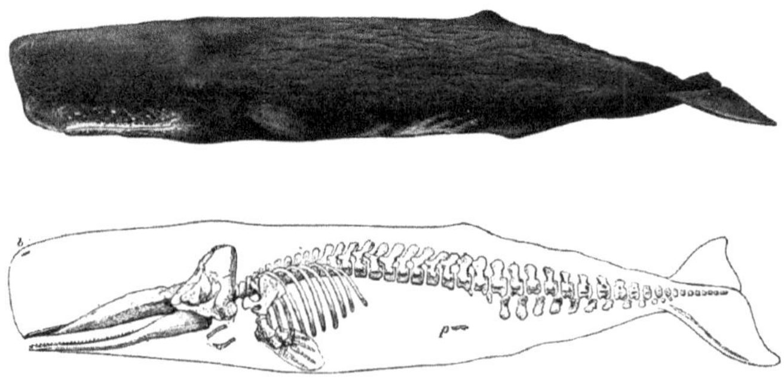

Sperm whale skeleton. Richard Lydekker, 1894.

The whale ear has specific adaptations to the marine environment. In humans, the middle ear works as an impedance equalizer between the outside air's low impedance and the cochlear fluid's high impedance. In whales, and other marine mammals, there is no great difference between the outer and inner environments. Instead of sound passing through the outer ear to the middle ear, whales receive sound through the throat, from which it passes through a low-impedance fat-filled cavity to the inner ear. The whale ear is acoustically isolated from the skull by air-filled sinus pockets, which allow for greater directional hearing underwater. Odontocetes send out high frequency clicks from an organ known as a melon. This melon consists of fat, and the skull of any such creature containing a melon will have a large depression. The melon size varies between species, the bigger the more dependent they are of it. A beaked whale for example has a small bulge sitting on top of its skull, whereas a sperm whale's head is filled up mainly with the melon.

The whale eye is relatively small for its size, yet they do retain a good degree of eyesight. As well as this, the eyes of a whale are placed on the sides of its head, so their vision consists of two fields, rather than a binocular view like humans have. When belugas surface, their lens and cornea correct the nearsightedness that results from the refraction of light; they contain both rod and cone cells, meaning they can see in both dim and bright light, but they have far more rod cells than they do cone cells. Whales do, however, lack short wavelength sensitive visual pigments in their cone cells indicating a more limited capacity for colour vision than most mammals. Most whales have slightly flattened eyeballs, enlarged pupils (which shrink as they surface to prevent damage), slightly flattened corneas and a tapetum lucidum; these adaptations allow for large amounts of light to pass through the eye and, therefore, a very clear image of the surrounding area. In water, a whale can see around 10.7 metres (35 ft) ahead of itself, but, of course, they have a smaller range above water. They also have glands on the eyelids and outer corneal layer that act as protection for the cornea.

The olfactory lobes are absent in toothed whales, suggesting that they have no sense of smell. Some whales, such as the bowhead whale, possess a vomeronasal organ, which does mean that they can "sniff out" krill. Whales are not thought to have a good sense of taste, as their taste buds are atrophied or missing altogether. However, some toothed whales have preferences between different kinds of fish, indicating some sort of attachment to taste. The presence of the Jacobson's organ indicates that whales can smell food once inside their mouth, which might be similar to the sensation of taste.

Communication

Whale vocalization is likely to serve several purposes. Some species, such as the humpback whale, communicate using melodic sounds, known as whale song. These sounds may be extremely loud, depending on the species. Humpback whales only have been heard making clicks, while toothed whales use sonar that may generate up to 20,000 watts of sound (+73 dBm or +43 dBw) and be heard for many miles. Captive whales have occasionally been known to mimic human speech. Scientists have suggested this indicates a strong desire on behalf of the whales to communicate with humans, as whales have a very different vocal mechanism, so imitating human speech likely takes considerable effort. Whales emit two distinct kinds of acoustic signals, which are called whistles and clicks: Clicks are quick broadband burst pulses, used for sonar, although some lower-frequency broadband vocalizations may serve a non-echolocative purpose such as communication; for example, the pulsed calls of belugas. Pulses in a click train are emitted at intervals of ~35–50 milliseconds, and in general these inter-click intervals are slightly greater than the round-trip time of sound to the target. Whistles are narrow-band frequency modulated (FM) signals, used for communicative purposes, such as contact calls.

Intelligence

Whales are known to teach, learn, cooperate, scheme, and grieve. The neocortex of many species of whale is home to elongated spindle neurons that, prior to 2007, were known only in hominids. In humans, these cells are involved in social conduct, emotions, judgement, and theory of mind. Whale spindle neurons are found in areas of the brain that are homologous to where they are found in humans, suggesting that they perform a similar function.

Bubble net feeding

Brain size was previously considered a major indicator of the intelligence of an animal. Since most of the brain is used for maintaining bodily functions, greater ratios of brain to body mass may increase the amount of brain mass available for more complex cognitive tasks. Allometric analysis indicates that mammalian brain size scales at approximately the ⅔ or ¾ exponent of the body mass. Comparison of a particular animal's brain size with the expected brain size based on such allometric analysis provides an encephalisation quotient that can be used as another indication of animal intelligence. Sperm whales have the largest brain mass of any animal on earth, averaging 8,000 cubic centimetres (490 in^3) and 7.8 kilograms (17 lb) in mature males, in comparison to the average human brain which averages 1,450 cubic centimetres (88 in^3) in mature males. The brain to body mass ratio in some odontocetes, such as belugas and narwhals, is second only to humans.

Small whales are known to engage in complex play behaviour, which includes such things as producing stable underwater toroidal air-core vortex rings or "bubble rings". There are two main methods of bubble ring production: rapid puffing of a burst of air into the water and allowing it to rise to the surface, forming a ring, or swimming repeatedly in a circle and then stopping to inject air into the helical vortex currents thus formed. They also appear to enjoy biting the vortex-rings, so that they burst into many separate bubbles and then rise quickly to the surface. Some believe this is a means of communication. Whales are also known to produce bubble-nets for the purpose of foraging.

A southern right whale sailing

Larger whales are also thought, to some degree, to engage in play. The southern right whale, for example, elevates their tail fluke above the water, remaining in the same position for a considerable amount of time. This is known as "sailing". It appears to be a form of play and is most commonly seen off the coast of Argentina and South Africa. Humpback whales, among others, are also known to display this behaviour.

Life Cycle

Whales are fully aquatic creatures, which means that birth and courtship behaviours are very different from terrestrial and semi-aquatic creatures. Since they are unable to go onto land to calve, they deliver the baby with the fetus positioned for tail-first delivery. This prevents the baby from drowning either upon or during delivery. To feed the new-born, whales, being aquatic, must squirt the milk into the mouth of the calf. Being mammals, they, of course, have mammary glands used for nursing calves; they are weaned off at about 11 months of age. This milk contains high amounts of fat which is meant to hasten the development of blubber; it contains so much fat that it has the consistency of toothpaste. Females deliver a single calf with gestation lasting about a year, dependency until one to two years, and maturity around seven to ten years, all varying between the species. This mode of reproduction produces few offspring, but increases the survival probability of each one. Females, referred to as "cows", carry the responsibility of childcare as males, referred to as "bulls", play no part in raising calves.

Most mysticetes reside at the poles. So, to prevent the unborn calf from dying of frostbite, they migrate to calving/mating grounds. They will then stay there for a matter of months until the calf has developed enough blubber to survive the bitter temperatures of the poles. Until then, the calves will feed on the mother's fatty milk. With the exception of the humpback whale, it is largely unknown when whales migrate. Most will travel from the Arctic or Antarctic into the tropics to mate, calve, and raise during the winter and spring; they will migrate back to the poles in the warmer summer months so the calf can continue growing while the mother can continue eating, as they fast in the breeding grounds. One exception to this is the southern right whale, which migrates to Patagonia and western New Zealand to calve; both are well out of the tropic zone.

Sleep

Unlike most animals, whales are conscious breathers. All mammals sleep, but whales cannot afford to become unconscious for long because they may drown. While knowledge of sleep in wild cetaceans is limited, toothed cetaceans in captivity have been recorded to sleep with one side of their brain at a time, so that they may swim, breathe consciously, and avoid both predators and social contact during their period of rest.

A 2008 study found that sperm whales sleep in vertical postures just under the surface in passive shallow 'drift-dives', generally during the day, during which whales do not respond to passing vessels unless they are in contact, leading to the suggestion that whales possibly sleep during such dives.

Ecology

Foraging and Predation

Polar bear with the remains of a beluga

All whales are carnivorous and predatory. Odontocetes, as a whole, mostly feed on fish and cephalopods, and then followed by crustaceans and bivalves. All species are generalist and opportunistic feeders. Mysticetes, as a whole, mostly feed on krill and plankton, followed by crustaceans and other invertebrates. A few are specialists. Examples include the blue whale, which eats almost exclusively krill, the minke whale, which eats mainly schooling fish, the sperm whale, which specialize on squid, and the gray whale which feed on bottom-dwelling invertebrates. The elaborate baleen "teeth" of filter-feeding species, mysticetes, allow them to remove water before they swallow their planktonic food by using the teeth as a sieve. Usually whales hunt solitarily, but they do sometimes hunt cooperatively in small groups. The former behaviour is typical when hunting non-schooling fish, slow-moving or immobile invertebrates or endothermic prey. When large amounts of prey are available, whales such as certain mysticetes hunt cooperatively in small groups. Some cetaceans may forage with other kinds of animals, such as other species of whales or certain species of pinnipeds.

Large whales, such as mysticetes, are not usually subject to predation, but smaller whales, such as monodontids or ziphiids, are. These species are preyed on by the killer whale or orca. To subdue and kill whales, orcas continuously ram them with their heads; this can sometimes kill bowhead whales, or severely injure them. Other times they corral the narwhals or belugas before striking. They are typically hunted by groups of 10 or fewer orcas, but they are seldom attacked by an individual. Calves are more commonly taken by orcas, but adults can be targeted as well.

These small whales are also targeted by terrestrial and pagophilic predators. The polar bear is well adapted for hunting Arctic whales and calves. Bears are known to use sit-and-wait tactics as well as active stalking and pursuit of prey on ice or water. Whales lessen the chance of predation by gathering in groups. This however means less room around the breathing hole as the ice slowly closes the gap. When out at sea, whales dive out of the reach of surface-hunting orcas. Polar bear attacks on belugas and narwhals are usually successful in winter, but rarely inflict any damage in summer.

Whale Pump

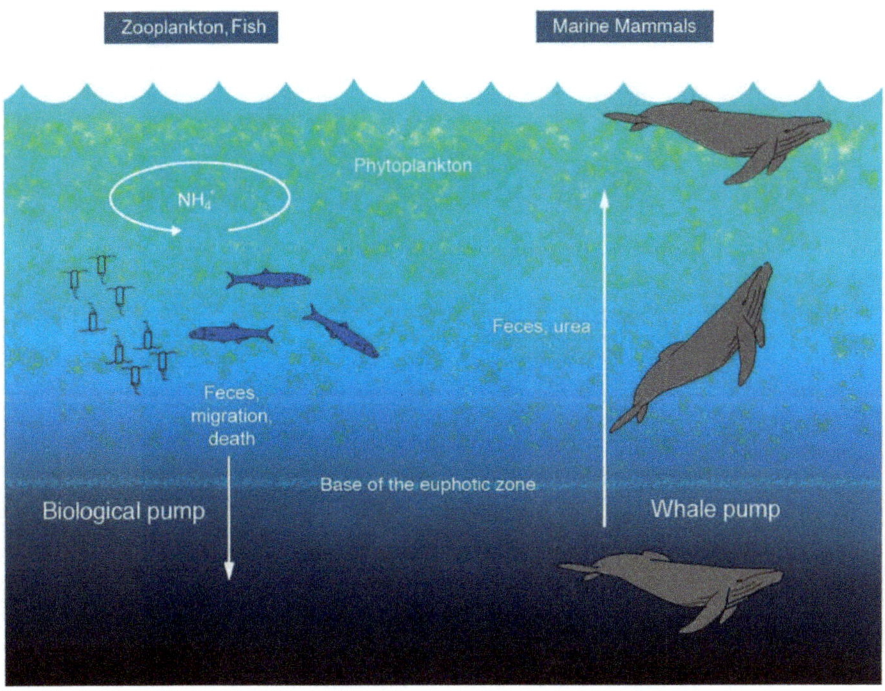

"Whale pump" – the role played by whales in recycling ocean nutrients

A 2010 study considered whales to be a positive influence to the productivity of ocean fisheries, in what has been termed a "whale pump." Whales carry nutrients such as nitrogen from the depths back to the surface. This functions as an upward biological pump, reversing an earlier presumption that whales accelerate the loss of nutrients to the bottom. This nitrogen input in the Gulf of Maine is "more than the input of all rivers combined" emptying into the gulf, some 23,000 metric tons (25,000 short tons) each year. Whales defecate at the ocean's surface; their excrement is important for fisheries because it is rich in iron and nitrogen. The whale faeces are liquid and instead of sinking, they stay at the surface where phytoplankton feed off it.

Whale Fall

Upon death, whale carcasses fall to the deep ocean and provide a substantial habitat for marine life. Evidence of whale falls in present-day and fossil records shows that deep sea whale falls support a rich assemblage of creatures, with a global diversity of 407 species, comparable to other neritic biodiversity hotspots, such as cold seeps and hydrothermal vents.

Deterioration of whale carcasses happens though a series of three stages. Initially, moving organisms such as sharks and hagfish, scavenge the soft tissues at a rapid rate over a period of months, and as long as two years. This is followed by the colonization of bones and surrounding sediments (which contain organic matter) by enrichment opportunists, such as crustaceans and polychaetes, throughout a period of years. Finally, sulfophilic bacteria reduce the bones releasing hydrogen sulphide enabling the growth of chemoautotrophic organisms, which in turn, support other organisms such as mussels, clams, limpets, and sea snails. This stage may last for decades and supports a rich assemblage of species, averaging 185 species per site.

Interaction with Humans

Whaling

Whale Fishing: Woodcut by Thevet, Paris, 1574

Dutch whalers near Spitsbergen, their most successful port. *Abraham Storck, 1690*

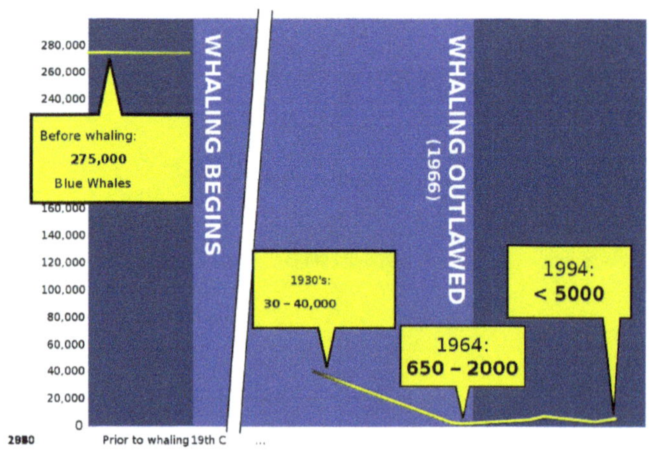

World population graph of blue whales

Whaling by humans has existed since the Stone Age. Ancient whalers used harpoons to spear the bigger animals from boats out at sea. People from Norway started hunting whales around 2000 B.C., and people from Japan began hunting whales in the Pacific at least as early. Whales are typically hunted for their meat and blubber by aboriginal groups; they used baleen for baskets or roofing, and made tools and masks out of bones. The Inuit hunted whales in the Arctic Ocean. The Basques started whaling as early as the 11th century, sailing as far as Newfoundland in the 16th century in search of right whales. 18th and 19th century whalers hunted down whales mainly for their oil, which was used as lamp fuel and a lubricant, baleen or whalebone, which was used for items such as corsets and skirt hoops, and ambergris, which was used as a fixative for perfumes. The most successful whaling nations at this time were the Netherlands, Japan, and the United States.

Commercial whaling was historically important as an industry well throughout the 17th, 18th and 19th centuries. Whaling was at that time a sizeable European industry with ships from Britain, France, Spain, Denmark, the Netherlands and Germany, sometimes collaborating to hunt whales in the Arctic, sometimes in competition leading even to war. By the early 1790s, whalers, namely the Americans and Australians, mainly focused efforts in the South Pacific where they mainly hunted sperm whales and right whales, with catches of up to 39,000 right whales by Americans alone. By 1853, U.S. profits turned to US$11,000,000 (UK£6.5m), equivalent to US$348,000,000 (UK£230m) today, the most profitable year for the American whaling industry. Commonly exploited species included North Atlantic right whales, sperm whales, which were mainly hunted by Americans, bowhead whales, which were mainly hunted by the Dutch, common minke whales, blue whales, and gray whales. The scale of whale harvesting decreased substantially after 1982 when the International Whaling Commission (IWC) placed a moratorium which set a catch limit for each country, excluding aboriginal groups until 2004.

Current whaling nations are Norway, Iceland, and Japan, despite their joining to the IWC, as well as the aboriginal communities of Siberia, Alaska, and northern Canada. Subsistence hunters typically use whale products for themselves and depend on them for survival. National and international authorities have given special treatment to aboriginal hunters since their methods of hunting are seen as less destructive and wasteful. This distinction is being questioned as these aboriginal groups are using more modern weaponry and mechanized transport to hunt with, and are selling whale products in the marketplace. Some anthropologists argue that the term "subsistence" should also apply to these cash-based exchanges as long as they take place within local production and consumption. In 1946, the IWC placed a moratorium, limiting the annual whale catch. Since then, yearly profits for these "subsistence" hunters have been close to US$31 million (UK£20m) per year.

Other Threats

Whales can also be threatened by humans more indirectly. They are unintentionally caught in fishing nets by commercial fisheries as bycatch and accidentally swallow fishing hooks. Gillnetting and Seine netting is a significant cause of mortality in whales and other marine mammals. Species commonly entangled include beaked whales. Whales are also affected by marine pollution. High levels of organic chemicals accumulate in these animals since they are high in the food chain. They have large reserves of blubber, more so for toothed whales as they are higher up the food chain than baleen whales. Lactating mothers can pass the toxins on to their young. These pollutants can cause gastrointestinal cancers and greater vulnerability to infectious diseases. They can also be poisoned by swallowing litter, such as plastic bags. Environmentalists speculate that advanced naval sonar endangers some whales. Some scientists suggest that sonar may trigger whale beachings, and they point to signs that such whales have experienced decompression sickness.

Conservation

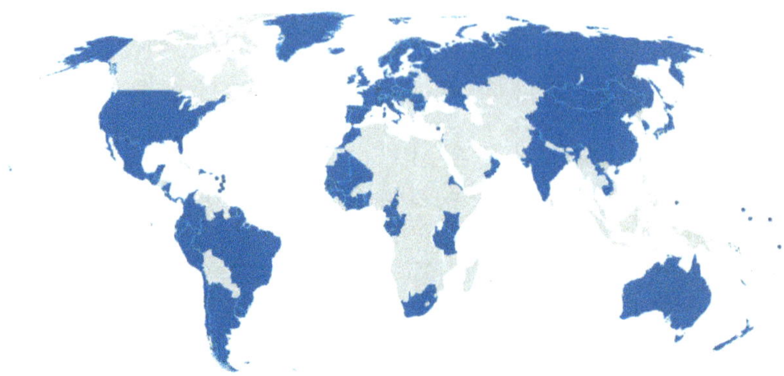

World map showing International Whaling Commission (IWC) members in blue

The scale of whale harvesting decreased substantially after 1946 when, in response to the steep decline in whale populations, the International Whaling Commission placed a moratorium which set a catch limit for each country; this excluded aboriginal groups up until 2004. As of 2015, aboriginal communities are allowed to take 280 bowhead whales off of Alaska and two from the western coast of Greenland, 620 gray whales off of Washington state, three common minke whales off of the eastern coast of Greenland and 178 on their western coast, 10 fin whales from the west coast of Greenland, nine humpback whales from the west coast of Greenland and 20 off of St. Vincent and the Grenadines each year. Several species that were commercially exploited have rebounded in numbers; for example, Grey whales may be as numerous as they were prior to harvesting, but the North Atlantic population is functionally extinct. Conversely, the North Atlantic right whale was extirpated from much of its former range, which stretched across the North Atlantic, and only remains in small fragments along the coast of Canada, Greenland, and is considered functionally extinct along the European coastline.

The IWC has designated two whale sanctuaries: the Southern Ocean Whale Sanctuary, and the Indian Ocean Whale Sanctuary. The Southern Ocean whale sanctuary spans 30,560,860 square kilometres (11,799,610 sq mi) and envelopes Antarctica. The Indian Ocean whale sanctuary takes up all of the Indian Ocean south of 55°S. The IWC is a voluntary organization, with no treaty. Any nation may leave as they wish; the IWC cannot enforce any law it makes.

As of 2013, the International Union for Conservation of Nature (IUCN) recognized 86 cetacean species, 40 of which are considered whales. Six are considered at risk, as they are ranked Critically Endangered (the North Atlantic right whale), "Endangered" (blue whale, fin whale, North Pacific right whale, and sei whale), and "Vulnerable" (sperm whale). Twenty-one species have a "Data Deficient" ranking. Species that live in polar habitats are vulnerable to the effects of recent and ongoing climate change, particularly the time when pack ice forms and melts.

Whale Watching

Whale watching off Bar Harbour, Maine

An estimated 13 million people went whale watching globally in 2008, in all oceans except the Arctic. Rules and codes of conduct have been created to minimize harassment of the whales. Iceland, Japan and Norway have both whaling and whale watching industries. Whale watching lobbyists are concerned that the most inquisitive whales, which approach boats closely and provide much of the entertainment on whale-watching trips, will be the first to be taken if whaling is resumed in the same areas. Whale watching generated US$2.1 billion (UK£1.4 billion) per annum in tourism revenue worldwide, employing around 13,000 workers. In contrast, the whaling industry, with the moratorium in place, generates US$31 million (UK£20 million) per year. The size and rapid growth of the industry has led to complex and continuing debates with the whaling industry about the best use of whales as a natural resource.

In Myth, Literature and Art

Engraving by William van der Gouwen depicting a stranded sperm whale being butchered on the Dutch coast, 1598

Whalers off Twofold Bay, New South Wales. Watercolour by Oswald Brierly, 1867

As marine creatures that reside in either the depths or the poles, humans knew very little about whales over the course of history; many feared or revered them. The Nords and various arctic tribes revered the whale as they were important pieces of their lives. In Inuit creation myths, when 'Big Raven', a deity in human form, found a stranded whale, he was told by the Great Spirit where to find special mushrooms that would give him the strength to drag the whale back to the sea and thus, return order to the world. In an Icelandic legend, a man threw a stone at a fin whale and hit the blowhole, causing the whale to burst. The man was told not to go to sea for twenty years, but during the nineteenth year he went fishing and a whale came and killed him.

Whales played a major part in shaping the art forms of many coastal civilizations, such as the Norse, with some dating to the Stone Age. Petroglyphs off a cliff face in Bangudae, South Korea show 300 depictions of various animals, a third of which are whales. Some show particular detail in which there are throat pleats, typical of rorquals. These petroglyphs show these people, of around 7,000 to 3,500 B.C.E. in South Korea, had a very high dependency on whales.

The Pacific Islanders and Australian Aborigines viewed whales as bringers of good and joy. One exception is French Polynesia, where, in many parts, cetaceans are met with great brutality.

In Vietnam and Ghana, among other places, whales hold a sense of divinity. They are so respected in their cultures that they occasionally hold funerals for beached whales, a throwback to Vietnam's ancient sea-based Austro-Asiatic culture.The god of the seas, according to Chinese folklore, was a large whale with human limbs.

Whales have also played a role in sacred texts such as the Bible. It mentions whales in Genesis 1:21, Job 7:12, and Ezekiel 32:2. The "leviathan" described at length in Job 41:1-34 is generally understood to refer to a whale. The "sea monsters" in Lamentations 4:3 have been taken by some to refer to marine mammals, in particular whales, although most modern versions use the word "jackals" instead. The story of Jonah being swallowed by a great fish is told both in the Qur'an and in the Bible. A medieval column capital sculpture depicting this was made in the 12th century in the abbey church in Mozac, France. The Old Testament contains the Book of Jonah and in the New Testament, Jesus mentions this story in Matthew 12:40.

In 1585, Alessandro Farnese, 1585, and Francois, Duke of Anjou, 1582, were greeted on his ceremonial entry into the port city of Antwerp by floats including "Neptune and the Whale", indicating at least the city's dependence on the sea for its wealth.

Whales continue to be prevalent in modern literature. For example, Herman Melville's *Moby Dick* features a "great white whale" as the main antagonist for Ahab, who eventually is killed by it. The whale is an albino sperm whale, considered by Melville to be the largest type of whale, and is partly based on the historically attested bull whale Mocha Dick. Rudyard Kipling's *Just So Stories* includes the story of "How the Whale got in his Throat". Niki Caro's film the *Whale Rider* has a Māori girl ride a whale in her journey to be a suitable heir to the chieftain-ship. Walt Disney's film *Pinocchio* features a giant whale named Monstro as the final antagonist. Alan Hovhaness' orchestra *And God Created Great Whales* including the recorded sounds of humpback and bowhead whales. Léo Ferré's song "Il n'y a plus rien" is an example of biomusic that begins and ends with recorded whale songs mixed with a symphonic orchestra and his voice.

In Captivity

Beluga whales and trainers in an aquarium

Belugas were the first whales to be kept in captivity. Other species were too rare, too shy, or too big. The first beluga was shown at Barnum's Museum in New York City in 1861. For most of the 20th century, Canada was the predominant source of wild belugas. They were taken from the St. Lawrence River estuary until the late 1960s, after which they were predominantly taken from the Churchill River estuary until capture was banned in 1992. Russia has become the largest provider since it had been banned in Canada. Belugas are caught in the Amur River delta and their eastern coast, and then are either transported domestically to aquariums or dolphinariums in Moscow, St. Petersburg, and Sochi, or exported to other countries, such as Canada.Most captive belugas are caught in the wild, since captive-breeding programs are not very successful.

As of 2006, 30 belugas were in Canada and 28 in the United States, and 42 deaths in captivity had been reported up to that time. A single specimen can reportedly fetch up to US$100,000 (UK£64,160) on the market. The beluga's popularity is due to its unique colour and its facial expressions. The latter is possible because while most cetacean "smiles" are fixed, the extra movement afforded by the beluga's unfused cervical vertebrae allows a greater range of apparent expression.

Between 1960 and 1992, the Navy carried out a program that included the study of marine mammals' abilities with sonar, with the objective of improving the detection of underwater objects. A large number of belugas were used from 1975 on, the first being dolphins. The program also included training them to carry equipment and material to divers working underwater by holding cameras in their mouths to locate lost objects, survey ships and submarines, and underwater monitoring. A similar program was used by the Russian Navy during the Cold War, in which belugas were also trained for antimining operations in the Arctic.

Aquariums have tried housing other species of whales in captivity. The success of belugas turned attention to maintaining their relative, the narwhal, in captivity. However, in repeated attempts in the 1960s and 1970s, all narwhals kept in captivity died within months. A pair of pygmy right whales were retained in an enclosed area (with nets); they were eventually released in South Africa. There was one attempt to keep a stranded Sowerby's beaked whale calf in captivity; the calf rammed into the tank wall, breaking its rostrum, which resulted in death. It was thought that Sowerby's beaked whale evolved to swim fast in a straight line, and a 30 metres (98 ft) was not big enough. There have been attempts to keep baleen whales in captivity. There were three attempts to keep gray whales in captivity. Gigi was a gray whale calf that died in transport. Gigi II was another gray whale calf that was captured in the Ojo de Liebre Lagoon, and was transported to SeaWorld. The 680 kilograms (1,500 lb) calf was a popular attraction, and behaved normally, despite being separated from his mother. A year later, the 8,000-kilogram (18,000 lb) whale grew too big to keep in captivity and was released; it was the first of two gray whales, the other being another gray whale calf named JJ, to successfully be kept in captivity. There were three attempts to keep minke whales in captivity in Japan. They were kept in a tidal pool with a sea-gate at the Izu Mito Sea Paradise. Another, unsuccessful, attempt was made by the U.S. One stranded humpback whale calf was kept in captivity for rehabilitation, but died days later.

DOLPHIN

Dolphins are a widely distributed and diverse group of aquatic mammals. They are an informal grouping within the order Cetacea, excluding whales and porpoises, so to zoologists the grouping is paraphyletic. The dolphins comprise the extant families Delphinidae (the oceanic dolphins), Platanistidae (the Indian river dolphins), Iniidae (the new world river dolphins), and Pontoporiidae (the brackish dolphins). There are 40 extant species of dolphins. Dolphins, alongside other cetaceans, belong to the clade Cetartiodactyla with even-toed ungulates. Their closest living relatives are the hippopotamuses, having diverged about 40 million years ago.

Dolphins range in size from the 1.7 m (5.6 ft) long and 50 kg (110 lb) Maui's dolphin to the 9.5 m (31 ft) and 10 t (11 short tons) killer whale. Several species exhibit sexual dimorphism, in that the males are larger than females. They have streamlined bodies and two limbs that are modified into flippers. Though not quite as flexible as seals, some dolphins can travel at 55.5 km/h (34.5 mph). Dolphins use their conical shaped teeth to capture fast moving prey. They have well-developed hearing—their hearing, which is adapted for both air and water, is so well developed that some can survive even if they are blind. Some species are well adapted for diving to great depths. They have a layer of fat, or blubber, under the skin to keep warm in the cold water.

Although dolphins are widespread, most species prefer the warmer waters of the tropic zones, but some, like the right whale dolphin, prefer colder climates. Dolphins feed largely on fish and squid, but a few, like the killer whale, feed on large mammals, like seals. Male dolphins typically mate with multiple females every year, but females only mate every two to three years. Calves are typically born in the spring and summer months and females bear all the responsibility for raising them. Mothers of some species fast and nurse their young for a relatively long period of time. Dolphins produce a variety of vocalizations, usually in the form of clicks and whistles.

Dolphins are sometimes hunted in places like Japan, in an activity known as dolphin drive hunting. Besides drive hunting, they also face threats from bycatch, habitat loss, and marine pollution. Dolphins have been depicted in various cultures worldwide. Dolphins occasionally feature in literature and film, as in the film series *Free Willy*. Dolphins are sometimes kept in captivity and trained to perform tricks, but breeding success has been poor and the animals often die within a few months of capture. The most common dolphins kept are killer whales and bottlenose dolphins.

Bottlenose dolphin

Etymology

The name is originally from Greek δελφίς (*delphís*), "dolphin", which was related to the Greek δελφύς (*delphus*), "womb".The animal's name can therefore be interpreted as meaning "a 'fish' with a womb". The name was transmitted via the Latin*delphinus* which in Medieval Latin became *dolfinus* and in Old French *daulphin*, which reintroduced the *ph* into the word. The term **mereswine** (that is, "sea pig") has also historically been used.

The term ‹dolphin' can be used to refer to, under the parvorder Odontoceti, all the species in the family Delphinidae (oceanic dolphins) and the river dolphin families Iniidae (South American river dolphins), Pontoporiidae (La Plata dolphin), Lipotidae (Yangtze river dolphin) and Platanistidae (Ganges river dolphin and Indus river dolphin). This term has often been misused in the US, mainly in the fishing industry, where all small cetaceans (dolphins and porpoises) are considered porpoises, while the fish *dorado* is called dolphin fish. In common usage the term 'whale' is used only for the larger cetacean species, while the smaller ones with a beaked or longer nose are considered 'dolphins'. The name 'dolphin' is used casually as a synonym for bottlenose dolphin, the most common and familiar species of dolphin. There are six species of dolphins commonly thought of as whales, collectively known as blackfish: the killer whale, the melon-headed whale, the pygmy killer whale, the false killer whale, and the two species of pilot whales, all of which are classified under the family Delphinidae and qualify as dolphins. Though the terms ‹dolphin' and ‹porpoise' are sometimes used interchangeably, porpoises are not considered dolphins and have different physical features such as a shorter beak and spade-shaped teeth; they also differ in their behavior. Porpoises belong to the family Phocoenidae and share a common ancestry with the Delphinidae.

A group of dolphins is called a "school" or a "pod". Male dolphins are called "bulls", females "cows" and young dolphins are called "calves".

Taxonomy

Common dolphin

Bottlenose dolphin

Spotted dolphin

Atlantic spotted dolphin

Commerson's dolphin

Dusky dolphin

Killer whale

Amazon river dolphin

- Parvorder Odontoceti, toothed whales
    - Family Platanistidae
        - Ganges and Indus river dolphin, *Platanista gangetica* with two subspecies
            - Ganges river dolphin (or Susu), *Platanista gangetica gangetica*
            - Indus river dolphin (or Bhulan), *Platanista gangetica minor*
    - Family Iniidae
        - Amazon river dolphin (or Boto), *Inia geoffrensis*
            - Orinoco river dolphin (the Orinoco subspecies), *Inia geoffrensis humboldtiana*
        - Araguaian river dolphin (Araguaian boto), *Inia Araguaiaensis*
        - Bolivian river dolphin, *Inia boliviensis*
    - Family Lipotidae
        - Baiji (or Chinese river dolphin), *Lipotes vexillifer* (possibly extinct, since December 2006)
    - Family Pontoporiidae
        - La Plata dolphin (or Franciscana), *Pontoporia blainvillei*
    - Family Delphinidae, oceanic dolphins
        - Genus *Delphinus*
            - Long-beaked common dolphin, *Delphinus capensis*
            - Short-beaked common dolphin, *Delphinus delphis*
        - Genus *Tursiops*
            - Common bottlenose dolphin, *Tursiops truncatus*
            - Indo-Pacific bottlenose dolphin, *Tursiops aduncus*
            - Burrunan dolphin, *Tursiops australis*, a newly discovered species from the sea around Melbourne in September 2011.
        - Genus *Lissodelphis*

    - Northern right whale dolphin, *Lissodelphis borealis*
    - Southern right whale dolphin, *Lissodelphis peronii*
- Genus *Sotalia*
    - Tucuxi, *Sotalia fluviatilis*
    - Costero, *Sotalia guianensis*
- Genus *Sousa*
    - Indo-Pacific humpback dolphin, *Sousa chinensis*
        - Chinese white dolphin (the Chinese variant), *Sousa chinensis chinensis*
    - Atlantic humpback dolphin, *Sousa teuszii*
- Genus *Stenella*
    - Atlantic spotted dolphin, *Stenella frontalis*
    - Clymene dolphin, *Stenella clymene*
    - Pantropical spotted dolphin, *Stenella attenuata*
    - Spinner dolphin, *Stenella longirostris*
    - Striped dolphin, *Stenella coeruleoalba*
- Genus *Steno*
    - Rough-toothed dolphin, *Steno bredanensis*
- Genus *Cephalorhynchus*
    - Chilean dolphin, *Cephalorhynchus eutropia*
    - Commerson's dolphin, *Cephalorhynchus commersonii*
    - Haviside's dolphin, *Cephalorhynchus heavisidii*
    - Hector's dolphin, *Cephalorhynchus hectori*
- Genus *Grampus*
    - Risso's dolphin, *Grampus griseus*
- Genus *Lagenodelphis*
    - Fraser's dolphin, *Lagenodelphis hosei*

- Genus *Lagenorhynchus*
    - Atlantic white-sided dolphin, *Lagenorhynchus acutus*
    - Dusky dolphin, *Lagenorhynchus obscurus*
    - Hourglass dolphin, *Lagenorhynchus cruciger*
    - Pacific white-sided dolphin, *Lagenorhynchus obliquidens*
    - Peale's dolphin, *Lagenorhynchus australis*
    - White-beaked dolphin, *Lagenorhynchus albirostris*
- Genus *Orcaella*
    - Australian snubfin dolphin, *Orcaella heinsohni*
    - Irrawaddy dolphin, *Orcaella brevirostris*
- Genus *Peponocephala*
    - Melon-headed whale, *Peponocephala electra*
- Genus *Orcinus*
    - Killer whale (Orca), *Orcinus orca*
- Genus *Feresa*
    - Pygmy killer whale, *Feresa attenuata*
- Genus *Pseudorca*
    - False killer whale, *Pseudorca crassidens*
- Genus *Globicephala*
    - Long-finned pilot whale, *Globicephala melas*
    - Short-finned pilot whale, *Globicephala macrorhynchus*
- Genus †Australodelphis
    - †*Australodelphis mirus*

- Six species in the family Delphinidae are commonly called "whales", but genetically are dolphins. They are sometimes called *blackfish*.
    - Melon-headed whale, *Peponocephala electra*
    - Killer whale (Orca), *Orcinus orca*

- Pygmy killer whale, *Feresa attenuata*
- False killer whale, *Pseudorca crassidens*
- Long-finned pilot whale, *Globicephala melas*
- Short-finned pilot whale, *Globicephala macrorhynchus*

Hybridization

In 1933, three strange dolphins beached off the Irish coast; they appeared to be hybrids between Risso's and bottlenose dolphins. This mating was later repeated in captivity, producing a hybrid calf. In captivity, a bottlenose and a rough-toothed dolphin produced hybrid offspring. A common-bottlenose hybrid lives at SeaWorld California. Other dolphin hybrids live in captivity around the world or have been reported in the wild, such as a bottlenose-Atlantic spotted hybrid. The best known hybrid is the wolphin, a false killer whale-bottlenose dolphin hybrid. The wolphin is a fertile hybrid. Two wolphins currently live at the Sea Life Park in Hawaii; the first was born in 1985 from a male false killer whale and a female bottlenose. Wolphins have also been observed in the wild.

Evolution

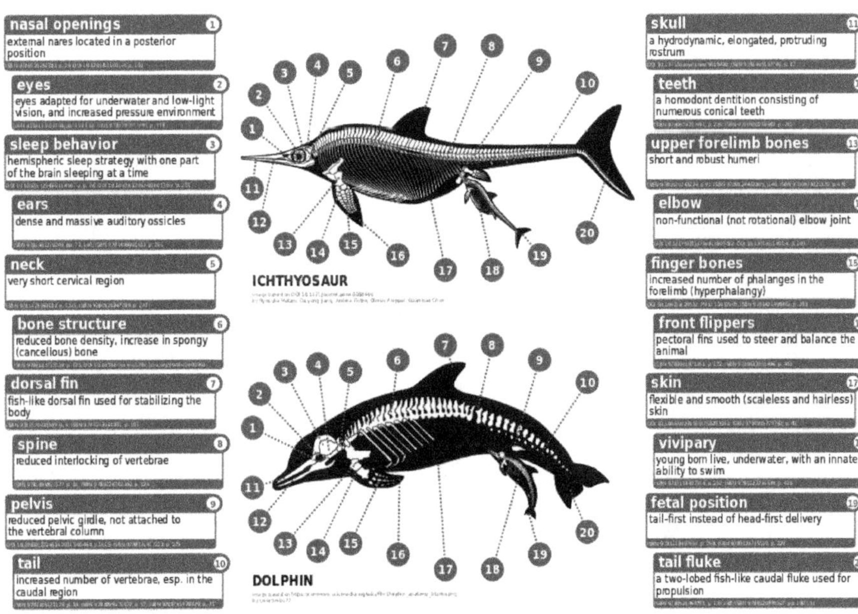

Dolphins display convergent evolution with fish and aquatic reptiles

Dolphins are descendants of land-dwelling mammals of the artiodactyl order (even-toed ungulates). They are related to the *Indohyus*, an extinct chevrotain-like ungulate, from which they split approximately 48 million years ago. The primitive cetaceans, or archaeocetes, first took to the sea approximately 49 million years ago and became fully aquatic by 5–10 million years later.

Archaeoceti is a parvorder comprising ancient whales. These ancient whales are the predecessors of modern whales, stretching back to their first ancestor that spent their lives near (rarely in) the water. Likewise, the archaeocetes can be anywhere from near fully terrestrial, to semi-aquatic to fully aquatic, but what defines an archaeocete is the presence of visible legs or asymmetrical teeth. Their features became adapted for living in the marine environment. Major anatomical changes include the hearing set-up that channeled vibrations from the jaw to the earbone which occurred with *Ambulocetus* 49 million years ago, a streamlining of the body and the growth of flukes on the tail which occurred around 43 million years ago with *Protocetus*, the migration of the nasal openings toward the top of the cranium and the modification of the forelimbs into flippers which occurred with *Basilosaurus* 35 million years ago, and the shrinking and eventual disappearance of the hind limbs which took place with the first odontocetes and mysticetes 34 million years ago. The modern dolphin skeleton has two small, rod-shaped pelvic bones thought to be vestigial hind limbs. In October 2006, an unusual bottlenose dolphin was captured in Japan; it had small fins on each side of its genital slit, which scientists believe to be an unusually pronounced development of these vestigial hind limbs.

Today, the closest living relatives of cetaceans are the hippopotamuses; these share a semi-aquatic ancestor that branched off from other artiodactyls some 60 million years ago. Around 40 million years ago, a common ancestor between the two branched off into cetacea and anthracotheres; anthracotheres became extinct at the end of the Pleistocene two-and-a-half million years ago, eventually leaving only one surviving lineage: the hippo.

Biology

Anatomy

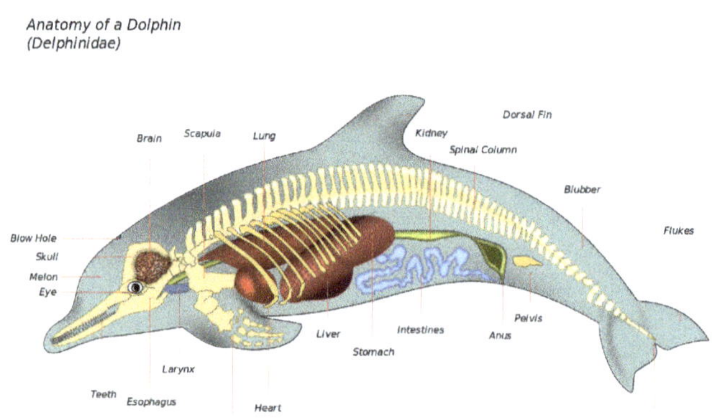

The anatomy of a dolphin showing its skeleton, major organs, tail and body shape

Dolphins have torpedo shaped bodies with generally non-flexible necks, limbs modified into flippers, non-existent external ear flaps, a tail fin, and bulbous heads. Dolphin skulls have small eye orbits, long snouts, and eyes placed on the sides of its head. Dolphins range in size from the 1.7 metres (5.6 ft) long and 50 kilograms (110 lb) Maui's dolphin to the 9.5 metres (31 ft) and 10 metric tons (11 short tons) killer whale. Overall, however, they tend to be dwarfed by other Cetartiodactyls. Several species have female-biased sexual dimorphism, with the females being larger than the males.

Dolphins have conical shape teeth, as apposed to their counterparts, porpoise's, spade-shaped teeth. These conical teeth are used to catch swift prey such as fish, squid or large mammals, such as seal.

Breathing involves expelling stale air from their blowhole, forming an upward, steamy spout, followed by inhaling fresh air into the lungs, however this only occurs in the polar regions of the oceans. Dolphins have rather small, unidentifiable spouts.

All dolphins have a thick layer of blubber, thickness varying on climate. This blubber can help with buoyancy, protection to some extent as predators would have a hard time getting through a thick layer of fat, and energy for leaner times; the primary usage for blubber is insulation from the harsh climate. Calves, generally, are born with a thin layer of blubber, which develops at different paces depending on the habitat.

Dolphins have a two-chambered stomach that is similar in structure to terrestrial carnivores. They have fundic and pyloric chambers.

Locomotion

Dolphins have two flippers on the underside toward the head, a dorsal fin and a tail fin. These flippers contain four digits. Although dolphins do not possess fully developed hind limbs, some possess discrete rudimentary appendages, which may contain feet and digits. Dolphins are fast swimmers in comparison to seals who typically cruise at 9–28 kilometres per hour (5.6–17.4 mph); the killer whale, in comparison, can travel at speeds up to 55.5 kilometres per hour (34.5 mph). The fusing of the neck vertebrae, while increasing stability when swimming at high speeds, decreases flexibility, which means they are unable to turn their heads. River dolphins, however, have non-fused neck vertebrae and are able to turn their head up to 90°. Dolphins swim by moving their tail fin and rear body vertically, while their flippers are mainly used for steering. Some species log out of the water, which may allow them to travel faster. Their skeletal anatomy allows them to be fast swimmers. All species have a dorsal fin to prevent themselves from involuntarily spinning in the water.

Some dolphins are adapted for diving to great depths. In addition to their streamlined bodies, some can slow their heart rate to conserve oxygen. Some can also re-route blood from tissue tolerant of water pressure to the heart, brain and other organs. Their hemoglobin and myoglobin store oxygen in body tissues and they have twice the concentration of myoglobin than hemoglobin.

Senses

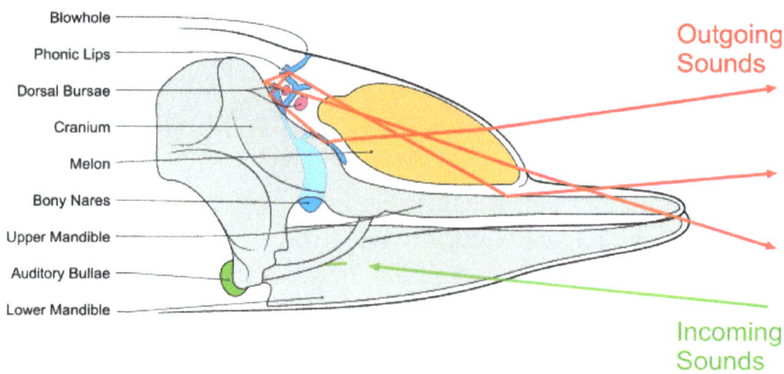

The dolphin ear has specific adaptations to the marine environment. In humans, the middle ear works as an impedance equalizer between the outside air's low impedance and the cochlear fluid's high impedance. In dolphins, and other marine mammals, there is no great difference between the outer and inner environments. Instead of sound passing through the outer ear to the middle ear, dolphins receive sound through the throat, from which it passes through a low-impedance fat-filled cavity to the inner ear. The dolphin ear is acoustically isolated from the skull by air-filled sinus pockets, which allow for greater directional hearing underwater. Dolphins send out high frequency clicks from an organ known as a melon. This melon consists of fat, and the skull of any such creature containing a melon will have a large depression. This allows dolphins to produce biosonar for orientation. Though most dolphins do not have hair, they do have hair follicles that may perform some sensory function. Beyond locating an object, echolocation also provides the animal with an idea on an object's shape and size, though how exactly this works is not yet understood. The small hairs on the rostrum of the boto are believed to function as a tactile sense, possibly to compensate for the boto's poor eyesight.

The dolphin eye is relatively small for its size, yet they do retain a good degree of eyesight. As well as this, the eyes of a dolphin are placed on the sides of its head, so their vision consists of two fields, rather than a binocular view like humans have. When dolphins surface, their lens and cornea correct the nearsightedness that results from the refraction of light; they contain both rod and cone cells, meaning they can see in both dim and bright light, but they have far more rod cells than they do cone cells. Dolphins do, however, lack short wavelength sensitive visual pigments in their cone cells indicating a more limited capacity for color vision than most mammals. Most dolphins have slightly flattened eyeballs, enlarged pupils (which shrink as they surface to prevent damage), slightly flattened corneas and a tapetum lucidum; these adaptations allow for large amounts of light to pass through the eye and, therefore, a very clear image of the surrounding area. They also have glands on the eyelids and outer corneal layer that act as protection for the cornea. The olfactory lobes are absent in dolphins, suggesting that they have no sense of smell.

Dolphins are not thought to have a good sense of taste, as their taste buds are atrophied or missing altogether. However, some have preferences between different kinds of fish, indicating some sort of attachment to taste.

Behavior

A pod of Indo-Pacific bottlenose dolphins in the Red Sea

Dolphins are often regarded as one of Earth's most intelligent animals, though it is hard to say just how intelligent. Comparing species' relative intelligence is complicated by differences in sensory apparatus, response modes, and nature of cognition. Furthermore, the difficulty and expense of experimental work with large aquatic animals has so far prevented some tests and limited sample size and rigor in others. Compared to many other species, however, dolphin behavior has been studied extensively, both in captivity and in the wild. See cetacean intelligence for more details.

Social Behavior

Dolphins surfing at Snapper Rocks, Queensland, Australia

Dolphins are highly social animals, often living in pods of up to a dozen individuals, though pod sizes and structures vary greatly between species and locations. In places with a high abundance of food, pods can merge temporarily, forming a *superpod*; such groupings may exceed 1,000 dolphins. Membership in pods is not rigid; interchange is common. Dolphins can, however, establish strong social bonds; they will stay with injured or ill individuals, even helping them to breathe by bringing them to the surface if needed. This altruism does not appear to be limited to their own species. The dolphin *Moko* in New Zealand has been observed guiding a female Pygmy Sperm Whale together with her calf out of shallow water where they had stranded several times. They have also been seen protecting swimmers from sharks by swimming circles around the swimmers or charging the sharks to make them go away.

Dolphins communicate using a variety of clicks, whistle-like sounds and other vocalizations. Dolphins also use nonverbal communication by means of touch and posturing.

Dolphins also display culture, something long believed to be unique to humans (and possibly other primate species). In May 2005, a discovery in Australia found Indo-Pacific bottlenose dolphins (*Tursiops aduncus*) teaching their young to use tools. They cover their snouts with sponges to protect them while foraging. This knowledge is mostly transferred by mothers to daughters, unlike simian primates, where knowledge is generally passed on to both sexes. Using sponges as mouth protection is a learned behavior. Another learned behavior was discovered among river dolphins in Brazil, where some male dolphins use weeds and sticks as part of a sexual display.

Forms of care-giving between fellows and even for members of different species (see Moko (dolphin)) are recorded in various species - such as trying to save weakened fellows or female pilot whales holding up dead calves for long periods.

Dolphins engage in acts of aggression towards each other. The older a male dolphin is, the more likely his body is to be covered with bite scars. Male dolphins engage in acts of aggression apparently for the same reasons as humans: disputes between companions and competition for females. Acts of aggression can become so intense that targeted dolphins sometimes go into exile after losing a fight.

Male bottlenose dolphins have been known to engage in infanticide. Dolphins have also been known to kill porpoises for reasons which are not fully understood, as porpoises generally do not share the same diet as dolphins and are therefore not competitors for food supplies.

Reproduction and Sexuality

Dolphins' reproductive organs are located on the underside of the body. Males have two slits, one concealing the penis and one further behind for the anus. The female has one genital slit, housing the vagina and the anus. Two mammary slits are positioned on either side of the female's genital slit.

Dolphin copulation happens belly to belly; though many species engage in lengthy foreplay, the actual act is usually brief, but may be repeated several times within a short timespan. The gestation period varies with species; for the small Tucuxi dolphin, this period is around 11 to 12 months, while for the orca, the gestation period is around

17 months. Typically dolphins give birth to a single calf, which is, unlike most other mammals, born tail first in most cases. They usually become sexually active at a young age, even before reaching sexual maturity. The age of sexual maturity varies by species and gender.

A skin-skeletal preparation

Dolphins are known to display non-reproductive sexual behavior, engaging in masturbation, stimulation of the genital area of other individuals using the rostrum or flippers, and homosexual contact.

Male dolphins have been known to masturbate by wrapping a live eel around their penis.

Various species of dolphin have been known to engage in sexual behavior up to and including copulation with dolphins of other species. Sexual encounters may be violent, with male dolphins sometimes showing aggressive behavior towards both females and other males. Male dolphins may also work together and attempt to herd females in estrus, keeping the females by their side by means of both physical aggression and intimidation, to increase their chances of reproductive success.Occasionally, dolphins behave sexually towards other animals, including humans.

Feeding

Various methods of feeding exist among and within species, some apparently exclusive to a single population. Fish and squid are the main food, but the false killer whale and the orca also feed on other marine mammals. Orcas on occasion also hunt whale species larger than themselves.

One common feeding method is herding, where a pod squeezes a school of fish into a small volume, known as a bait ball. Individual members then take turns plowing through the ball, feeding on the stunned fish. Coralling is a method where dolphins chase fish into shallow water to catch them more easily. Orcas and bottlenose dolphins have also been known to drive their prey onto a beach to feed on it, a behaviour known as beach or strand feeding. Some species also whack fish with their flukes, stunning them and sometimes knocking them out of the water. Reports of cooperative human-dolphin fishing date back to the ancient Roman author and natural philosopher Pliny the Elder. A modern human-dolphin partnership currently operates in Laguna, Santa Catarina, Brazil. Here, dolphins drive fish towards fishermen waiting along the shore and signal the men to cast their nets. The dolphins' reward is the fish that escape the nets.

Vocalizations

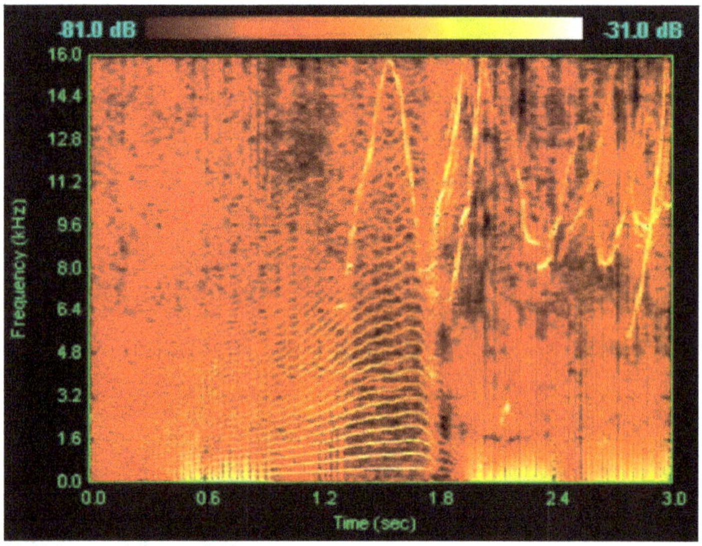

Spectrogram of dolphin vocalizations. Whistles, whines, and clicks are visible as upside down V's, horizontal striations, and vertical lines, respectively.

Dolphins are capable of making a broad range of sounds using nasal airsacs located just below the blowhole. Roughly three categories of sounds can be identified: frequency modulated whistles, burst-pulsed sounds and clicks. Dolphins communicate with whistle-like sounds produced by vibrating connective tissue, similar to the way human vocal cords function, and through burst-pulsed sounds, though the nature and extent of that ability is not known. The clicks are directional and are for echolocation, often occurring in a short series called a click train. The click rate increases when approaching an object of interest. Dolphin echolocation clicks are amongst the loudest sounds made by marine animals.

Bottlenose dolphins have been found to have signature whistles, a whistle that is unique to a specific individual. These whistles are used in order for dolphins to communicate with one another by identifying an individual. It can be seen as the dolphin equivalent of a name for humans. These signature whistles are developed during a dolphin's first year; it continues to maintain the same sound throughout its lifetime. In order to obtain each individual whistle sound, dolphins undergo vocal production learning. This consists of an experience with other dolphins that modifies the signal structure of an existing whistle sound. An auditory experience influences the whistle development of each dolphin. Dolphins are able to communicate to one another by addressing another dolphin through mimicking their whistle. The signature whistle of a male bottlenose dolphin tends to be similar to that of his mother, while the signature whistle of a female bottlenose dolphin tends to be more distinguishing. Bottlenose dolphins have a strong memory when it comes to these signature whistles, as they are able to relate to a signature whistle of an individual they have not encountered for over twenty years. Research done on signature whistle usage by other dolphin species is relatively limited. The research on other species done so far has yielded varied outcomes and inconclusive results.

Because dolphins are generally associated in groups, communication is necessary. Signal masking is when other similar sounds (conspecific sounds) interfere with the original acoustic sound. In larger groups, individual whistle sounds are less prominent. Dolphins tend to travel in pods, upon which there are groups of dolphins that range from a few to many. Although they are traveling in these pods, the dolphins do not necessarily swim right next to each other. Rather, they swim within the same general vicinity. In order to prevent losing one of their pod members, there are higher whistle rates. Because their group members were spread out, this was done in order to continue traveling together.

Jumping and Playing

Pacific white-sided dolphins porpoising

Dolphins frequently leap above the water surface, this being done for various reasons. When travelling, jumping can save the dolphin energy as there is less friction while in the air. This type of travel is known as porpoising. Other reasons include orientation, social displays, fighting, non-verbal communication, entertainment and attempting to dislodge parasites.

Dolphins show various types of playful behavior, often including objects, self-made bubble rings, other dolphins or other animals. When playing with objects or small animals, common behavior includes carrying the object or animal along using various parts of the body, passing it along to other members of the group or taking it from another member, or throwing it out of the water. Dolphins have also been observed harassing animals in other ways, for example by dragging birds underwater without showing any intent to eat them. Playful behaviour that involves another animal species with active participation of the other animal can also be observed however. Playful human interaction with dolphins being the most obvious example, however playful interactions have been observed in the wild with a number of other species as well, such as humpback whales and dogs.

Intelligence

Dolphins are known to teach, learn, cooperate, scheme, and grieve. The neocortex of many species is home to elongated spindle neurons that, prior to 2007, were known only in hominids. In humans, these cells are involved in social conduct, emotions, judgment, and theory of mind. Cetacean spindle neurons are found in areas of the brain that are homologous to where they are found in humans, suggesting that they perform a similar function.

Brain size was previously considered a major indicator of the intelligence of an animal. Since most of the brain is used for maintaining bodily functions, greater ratios of brain to body mass may increase the amount of brain mass available for more complex cognitive tasks. Allometric analysis indicates that mammalian brain size scales at approximately the ⅔ or ¾ exponent of the body mass. Comparison of a particular animal's brain size with the expected brain size based on such allometric analysis provides an encephalization quotient that can be used as another indication of animal intelligence. Killer whales have the second largest brain mass of any animal on earth, next to the sperm whale. The brain to body mass ratio in some is second only to humans.

Self-awareness is seen, by some, to be a sign of highly developed, abstract thinking. Self-awareness, though not well-defined scientifically, is believed to be the precursor to more advanced processes like meta-cognitive reasoning (thinking about thinking) that are typical of humans. Research in this field has suggested that cetaceans, among others, possess self-awareness. The most widely used test for self-awareness in animals is the mirror test in which a temporary dye is placed on an animal's body, and the animal is then presented with a mirror; they then see if the animal shows signs of self-recognition.

Some disagree with these findings, arguing that the results of these tests are open to human interpretation and susceptible to the Clever Hans effect. This test is much less definitive than when used for primates, because primates can touch the mark or the mirror, while cetaceans cannot, making their alleged self-recognition behavior less certain. Skeptics argue that behaviors that are said to identify self-awareness resemble existing social behaviors, and so researchers could be misinterpreting self-awareness for social responses to another individual. The researchers counter-argue that the behaviors shown are evidence of self-awareness, as they are very different from normal responses to another individual. Whereas apes can merely touch the mark on themselves with their fingers, cetaceans show less definitive behavior of self-awareness; they can only twist and turn themselves to observe the mark. In 1995, Marten and Psarakos used television to test dolphin self-awareness. They showed dolphins real-time footage of themselves, recorded footage, and another dolphin. They concluded that their evidence suggested self-awareness rather than social behavior. While this particular study has not been repeated since then, dolphins have since passed the mirror test.

Sleeping

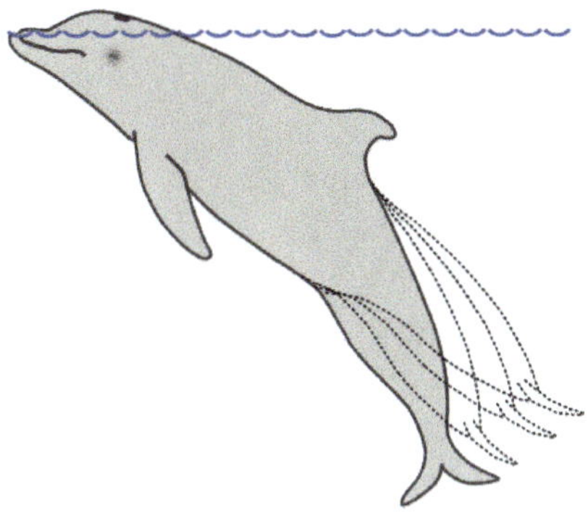

Sleeping dolphin in captivity: a tail kick reflex keeps the dolphin's blowhole above the water

Generally, dolphins sleep with only one brain hemisphere in slow-wave sleep at a time, thus maintaining enough consciousness to breathe and to watch for possible predators and other threats. Earlier sleep stages can occur simultaneously in both hemispheres. In captivity, dolphins seemingly enter a fully asleep state where both eyes are closed and there is no response to mild external stimuli. In this case, respiration is automatic; a tail kick reflex keeps the blowhole above the water if necessary. Anesthetized dolphins initially show a tail kick reflex. Though a similar state has been observed with wild sperm whales, it is not known if dolphins in the wild reach this state. The Indus river dolphin has a sleep method that is different from that of other dolphin species. Living in water with strong currents and potentially dangerous floating debris, it must swim continuously to avoid injury. As a result, this species sleeps in very short bursts which last between 4 and 60 seconds.

Threats

Lesions in the dorsal fin of a bottlenose dolphin caused by lobomycosis, a fungal infection of the skin

Natural Threats

Dolphins have few natural enemies. Some species or specific populations have none, making them apex predators. For most of the smaller species of dolphins, only a few of the larger sharks, such as the bull shark, dusky shark, tiger shark and great white shark, are a potential risk, especially for calves. Some of the larger dolphin species, especially orcas (killer whales), may also prey on smaller dolphins, but this seems rare. Dolphins also suffer from a wide variety of diseases and parasites. The Cetacean morbillivirus in particular has been known to cause regional epizootics often leaving hundreds of animals of various species dead.Symptoms of infection are often a severe combination of pneumonia, encephalitis and damage to the immune system, which greatly impair the cetacean's ability to swim and stay afloat unassisted. A study at the U.S. National Marine Mammal Foundation revealed that dolphins, like humans, develop a natural form of type 2 diabetes which may lead to a better understanding of the disease and new treatments for both humans and dolphins.

Dolphins can tolerate and recover from extreme injuries such as shark bites although the exact methods used to achieve this are not known. The healing process is rapid and even very deep wounds do not cause dolphins to hemorrhage to death. Furthermore, even gaping wounds restore in such a way that the animal's body shape is restored, and infection of such large wounds seems rare.

The study, published in the journal Marine Mammal Science, suggests that at least some dolphins are up to the shark challenge with research finding that many dolphins survive attacks using everything from sophisticated combat moves to teaming up against the shark.

Human Threats

Dead Atlantic white-sided dolphins in Hvalba on the Faroe Islands, killed in a drive hunt

Some dolphin species face an uncertain future, especially some river dolphin species such as the Amazon river dolphin, and the Ganges and Yangtze river dolphin, which are critically or seriously endangered. A 2006 survey found no individuals of the Yangtze river dolphin, which now appears to be functionally extinct.

Pesticides, heavy metals, plastics, and other industrial and agricultural pollutants that do not disintegrate rapidly in the environment concentrate in predators such as dolphins. Injuries or deaths due to collisions with boats, especially their propellers, are also common.

Various fishing methods, most notably purse seine fishing for tuna and the use of drift and gill nets, unintentionally kill many dolphins. Accidental by-catch in gill nets and incidental captures in antipredator nets that protect marine fish farms are common and pose a risk for mainly local dolphin populations. In some parts of the world, such as Taiji in Japan and the Faroe Islands, dolphins are traditionally considered food and are killed in harpoon or drive hunts. Dolphin meat is high in mercury and may thus pose a health danger to humans when consumed.

Dolphin safe labels attempt to reassure consumers that fish and other marine products have been caught in a dolphin-friendly way. The earliest campaigns with "Dolphin safe" labels were initiated in the 1980s as a result of cooperation between marine activists and the major tuna companies, and involved decreasing incidental dolphin kills by up to 50% by changing the type of nets used to catch tuna. The dolphins are netted only while fishermen are in pursuit of smaller tuna. Albacore are not netted this way, making albacore the only truly dolphin-safe tuna. Loud underwater noises, such as those resulting from navalsonar use, live firing exercises, and certain offshore construction projects such as wind farms, may be harmful to dolphins, increasing stress, damaging hearing, and causing decompression sickness by forcing them to surface too quickly to escape the noise.

Dolphins and other smaller cetaceans are also hunted in an activity known as dolphin drive hunting. This is accomplished by driving a pod together with boats and usually into a bay or onto a beach. Their escape is prevented by closing off the route to the ocean with other boats or nets. Dolphins are hunted this way in several places around the world, including the Solomon Islands, the Faroe Islands, Peru, and Japan, the most well-known practitioner of this method. By numbers, dolphins are mostly hunted for their meat, though some end up in dolphinariums. Despite the controversial nature of the hunt resulting in international criticism, and the possible health risk that the often polluted meat causes, thousands of dolphins are caught in drive hunts each year.

Relationships with Humans

In history and Religion

Fresco of Dolphins, ca. 1600 BC, from Knossos, Crete.

Vessel in form of killer whale, Nazca culture, circa 200 AD. American Museum of Natural History collections.

Dolphins have long played a role in human culture. Dolphins are sometimes used as symbols, for instance in heraldry. When heraldry developed in the Middle Ages, not much was known about the biology of the dolphin and it was often depicted as a sort of fish. Traditionally, the stylised dolphins in heraldry still may take after this notion, sometimes showing the dolphin skin covered with fish scales.

Dolphins are present in the coat of arms of Anguilla and the coat of arms of Romania, and the coat of arms of Barbados has a dolphin supporter. A well-known historical example of a dolphin in heraldry, was the arms for the former province of the Dauphiné in southern France, from which were derived the arms and the title of the Dauphin of France, the heir to the former throne of France (the title literally means "The Dolphin of France").

"Dolfin" was the name of an aristocratic family in the maritime Republic of Venice, whose most prominent member was the 13th Century Doge Giovanni Dolfin.

In Greek myths, they were seen invariably as helpers of humankind. Dolphins also seem to have been important to the Minoans, judging by artistic evidence from the ruined palace at Knossos. Dolphins are common in Greek mythology, and many coins from ancient Greece have been found which feature a man, a boy or a deity riding on the back of a dolphin. The Ancient Greeks welcomed dolphins; spotting dolphins riding in a ship's wake was considered a good omen. In both ancient and later art, Cupid is often shown riding a dolphin. A dolphin rescued the poet Arion from drowning and carried him safe to land, at Cape Matapan, a promontory forming the southernmost point of the Peloponnesus. There was a temple to Poseidon and a statue of Arion riding the dolphin. The Greeks reimagined the Phoenician god Melqart as Melikertês (Melicertes) and made him the son of Athamas and Ino. He drowned but was transfigured as the marine deity Palaemon, while his mother became Leucothea. (*cf* Ino.) At Corinth, he was so closely connected with the cult of Poseidon that the Isthmian Games, originally instituted in Poseidon's honor, came to be looked upon as the funeral games of Melicertes. Phalanthus was another legendary character brought safely to shore (in Italy) on the back of a dolphin, according to Pausanias.

Dionysus was once captured by Etruscan pirates who mistook him for a wealthy prince they could ransom. After the ship set sail Dionysus invoked his divine powers, causing vines to overgrow the ship where the mast and sails had been. He turned the oars into serpents, so terrifying the sailors that they jumped overboard, but Dionysus took pity on them and transformed them into dolphins so that they would spend their lives providing help for those in need. Dolphins were also the messengers of Poseidon and sometimes did errands for him as well. Dolphins were sacred to both Aphrodite and Apollo.

In Hindu mythology the Ganges River Dolphin is associated with Ganga, the deity of the Ganges river. The dolphin is said to be among the creatures which heralded the goddess' descent from the heavens and her mount, the Makara, is sometimes depicted as a dolphin.

The Boto, a species of river dolphin that resides in the Amazon River, are believed to be shapeshifters, or *encantados*, who are capable of having children with human women.

In Captivity

Species

SeaWorld show featuring bottlenose dolphins and pilot whales

The renewed popularity of dolphins in the 1960s resulted in the appearance of many dolphinaria around the world, making dolphins accessible to the public. Criticism and animal welfare laws forced many to close, although hundreds still exist around the world. In the United States, the best known are the SeaWorld marine mammal parks. In the Middle East the best known are Dolphin Bay at Atlantis, The Palm and the Dubai Dolphinarium.

SeaWorld Pilot Whale with trainers

Various species of dolphins are kept in captivity. These small cetaceans are more often than not kept in theme parks, such as SeaWorld, commonly known as a dolphinarium. Bottlenose Dolphins are the most common species of dolphin kept in dolphinariums as they are relatively easy to train, have a long lifespan in captivity and have a friendly appearance. Hundreds if not thousands of Bottlenose Dolphins live in captivity across the world, though exact numbers are hard to determine. Other species kept in captivity are Spotted Dolphins, False Killer Whales and Common Dolphins, Commerson's Dolphins, as well as Rough-toothed Dolphins, but all in much lower numbers than the Bottlenose Dolphin. There are also fewer than ten Pilot Whales, Amazon River Dolphins, Risso's Dolphins, Spinner Dolphins, or Tucuxi in captivity.An unusual and very rare hybrid dolphin, known as a Wolphin, is kept at the Sea Life Park in Hawaii, which is a cross between a Bottlenose Dolphin and a False Killer Whale.

Killer whales are well known for their performances in shows, but the number of Orcas kept in captivity is very small, especially when compared to the number of bottlenose dolphins, with only 44 captive killer whales being held in aquaria as of 2012. The killer whale's intelligence, trainability, striking appearance, playfulness in captivity and sheer size have made it a popular exhibit at aquaria and aquatic theme parks. From 1976 to 1997, 55 whales were taken from the wild in Iceland, 19 from Japan, and three from Argentina. These figures exclude animals that died during capture. Live captures fell dramatically in the 1990s, and by 1999, about 40% of the 48 animals on display in the world were captive-born.

Organizations such as the Mote Marine Laboratory rescue and rehabilitate sick, wounded, stranded or orphaned dolphins while others, such as the Whale and Dolphin Conservation Society and Hong Kong Dolphin Conservation Society, work on dolphin conservation and welfare. India has declared the dolphin as its national aquatic animal in an attempt to protect the endangered Ganges River Dolphin. The Vikramshila Gangetic Dolphin Sanctuary has been created in the Ganges river for the protection of the animals.

Controversy

Shamu the killer whale, 2009, with a collapsed dorsal fin

Organizations such as World Animal Protection and the Whale and Dolphin Conservation Society campaign against the practice of keeping them in captivity. In captivity, they often develop pathologies, such as the dorsal fin collapse seen in 60–90% of male killer whales. Captives have vastly reduced life expectancies, on average only living into their 20s, although there are examples of killer whales living longer, including several over 30 years old, and two captive orcas, Corky II and Lolita, are in their mid-40s. In the wild, females who survive infancy live 46 years on average, and up to 70–80 years in rare cases. Wild males who survive infancy live 31 years on average, and up to 50–60 years. Captivity usually bears little resemblance to wild habitat, and captive whales' social groups are foreign to those found in the wild. Critics claim captive life is stressful due to these factors and the requirement to perform circus tricks that are not part of wild killer whale behavior. Wild killer whales may travel up to 160 kilometres (100 mi) in a day, and critics say the animals are too big and intelligent to be suitable for captivity. Captives occasionally act aggressively towards themselves, their tankmates, or humans, which critics say is a result of stress.

Although dolphins generally interact well with humans, some attacks have occurred, most of them resulting in small injuries. Orcas, the largest species of dolphin, have been involved in fatal attacks on humans in captivity. The record-holder of documented orca fatal attacks is a male named Tilikum, who has lived at SeaWorld since 1992. Tilikum has played a role in the death of three people in three different incidents (1991, 1999 and 2010). Tilikum's behaviour sparked the production of the documentary *Blackfish*, which focuses on the consequences of keeping orcas in captivity. There are documented incidents in the wild, too, but none of them fatal.

Fatal attacks from other species are less common, but there is a registered occurrence off the coast of Brazil in 1994, when a man died after being attacked by a bottlenose dolphin named Tião. Tião had suffered harassment by human visitors, including attempts to stick ice cream sticks down her blowhole. Non-fatal incidents occur more frequently, both in the wild and in captivity.

While dolphin attacks occur far less frequently than attacks by other sea animals, such as sharks, some scientists are worried about the careless programs of human-dolphin interaction. Dr. Andrew J. Read, a biologist at the Duke University Marine Laboratory who studies dolphin attacks, points out that dolphins are large and wild predators, so people should be more careful when they interact with them.

Several scientists who have researched dolphin behaviour have proposed that dolphins' unusually high intelligence in comparison to other animals means that dolphins should be seen as non-human persons who should have their own specific rights and that it is morally unacceptable to keep them captive for entertainment purposes or to kill them either intentionally for consumption or unintentionally as by-catch. Four countries – Chile, Costa Rica, Hungary, and India – have declared dolphins to be "non-human persons" and have banned the capture and import of live dolphins for entertainment.

Military

A military dolphin

A military dolphin is a dolphin trained for military uses. A number of militaries have employed dolphins for various purposes from finding mines to rescuing lost or trapped humans. The military use of dolphins, however, drew scrutiny during the Vietnam War when rumors circulated that the United States Navy was training dolphins to kill Vietnamese divers. The United States Navy denies that at any point dolphins were trained for combat. Dolphins are still being trained by the United States Navy for other tasks as part of the U.S. Navy Marine Mammal Program. The Russian military is believed to have closed its marine mammal program in the early 1990s. In 2000 the press reported that dolphins trained to kill by the Soviet Navy had been sold to Iran.

Therapy

Dolphins are an increasingly popular choice of animal-assisted therapy for psychological problems and developmental disabilities. For example, a 2005 study found dolphins an effective treatment for mild to moderate depression. However, this study was criticized on several grounds. For example, it is not known whether dolphins are more effective than common pets. Reviews of this and other published dolphin-assisted therapy (DAT) studies have found important methodological flaws and have concluded that there is no compelling scientific evidence that DAT is a legitimate therapy or that it affords more than fleeting mood improvement.

Consumption

Cuisine

Plate of dolphin sashimi

In some parts of the world, such as Taiji, Japan and the Faroe Islands, dolphins are traditionally considered as food, and are killed in harpoon or drive hunts. Dolphin meat is consumed in a small number of countries worldwide, which include Japanand Peru (where it is referred to as *chancho marino*, or "sea pork"). While Japan may be the best-known and most controversial example, only a very small minority of the population has ever sampled it.

Dolphin meat is dense and such a dark shade of red as to appear black. Fat is located in a layer of blubber between the meat and the skin. When dolphin meat is eaten in Japan, it is often cut into thin strips and eaten raw as *sashimi*, garnished with onion and either horseradish or grated garlic, much as with *sashimi* of whale or horse meat (*basashi*). When cooked, dolphin meat is cut into bite-size cubes and then batter-fried or simmered in a *miso* sauce with vegetables. Cooked dolphin meat has a flavor very similar to beef liver.

Health Concerns

There have been human health concerns associated with the consumption of dolphin meat in Japan after tests showed that dolphin meat contained high levels of mercury. There are no known cases of mercury poisoning as a result of consuming dolphin meat, though the government continues to monitor people in areas where dolphin meat consumption is high. The Japanese government recommends that children and pregnant women avoid eating dolphin meat on a regular basis.

Similar concerns exist with the consumption of dolphin meat in the Faroe Islands, where prenatal exposure to methylmercury and PCBs primarily from the consumption of pilot whale meat has resulted in neuropsychological deficits amongst children.

Observations about Cetacea have been recorded since at least classical times. Ancient Greek fishermen created an artificial notch on the dorsal fin of dolphins entangled in nets so that they could tell them apart years later.

Approximately 2,300 years ago, Aristotle carefully took notes on cetaceans while traveling on boats with fishermen in the Aegean Sea. In his book *Historia animalium* (History of animals), Aristotle was careful enough to distinguish between the baleen whales and toothed whales, a taxonomical separation still used today. He also described the Sperm Whale and the common dolphin, stating that they can live for at least twenty-five or thirty years. His achievement was remarkable for its time, because even today it is very difficult to estimate the life-span of advanced marine animals. After Aristotle's death, much of the knowledge he had gained about cetaceans was lost, only to be re-discovered during the Renaissance.

Many of the medieval texts on cetaceans come mainly from Scandinavia and Iceland, most came about the mid-13th century. One of the better known is *Speculum Regale*. In this text is described various species that lived around the island of Iceland. It mentions "orcs" that had dog-like teeth and would demonstrate the same kind of aggression towards other cetaceans as wild dogs would to other terrestrial animals. The text even illustrated the hunting technique of Orcs, which are now called Orcas. The *Speculum Regale* describes other cetaceans, including the Sperm Whale and Narwhal. Many times they were seen as terrible monsters, such as killers of men, and destroyers of ships. They even bore odd names such as “Pig Whale”, “Horse Whale”, and “Red Whale”. But not all creatures described were said to be fierce. Some were seen to be good, such as whales that drove shoals of herring towards the shore. This was seen as very helpful to fisherman.

Many of the early studies were based on dead specimens and myth. The little information that was gathered was usually length, and a rough outer body anatomy. Because these animals live in water their entire lives, early scientists did not have the technology to go study these animals further. It was not until the 16th century that things would begin to change. Then cetaceans would be proved to be mammals rather than fish.

Aristotle argued they were mammals. But Pliny the Elder stated that they were fish, and it was followed by many naturalists. However, Pierre Belon (1517–1575) and G. Rondelet (1507–1566) persisted on convincing they were mammals. They argued that the animals had lungs and a uterus, just like mammals. Not until 1758, when Swedish botanist Carl Linnaeus (1707–1778) published the tenth edition of *Systema Naturae*, were they seen as mammals.

Only decades later, French zoologist and paleontologist Baron Georges Cuvier (1769–1832) described the animals as mammals without any hind legs. Skeletons were assembled and displayed in the first natural history museums, and on a closer look and comparisons with other extinct animal fossils led zoologists to conclude that cetaceans came from a family of ancient land mammals.

Between the 16th–20th century, much of our information on cetaceans came from whalers. Whalers were the most knowledgeable about the animals, but their information was regarding migration routes and outer anatomy, and only little information of behavior. During the 1960s, people began studying the animals intensively, often in dedicated research institutes. The Tethys Institute of Milan, founded in 1986, compiled an extensive cetology database of the Mediterranean. This came from both concern about wild populations and also the capture of larger animals such as the Orca, and gaining popularity of dolphin shows in marine parks.

Studying Cetaceans

Humpback Whales often have distinct markings that enable scientists to identify individuals.

Studying cetaceans presents numerous challenges. Cetaceans only spend 10% of their time on the surface, and all they do at the surface is breathe. There is very little behavior seen at the surface. It is also impossible to find any signs that an animal has been in an area. Cetaceans do not leave tracks that can be followed. However, the dung of whales often floats and can be collected to tell important information about their diet and about the role they have in the environment. Many times cetology consists of waiting and paying close attention.

Cetologists use equipment including hydrophones to listen to calls of communicating animals, binoculars and other optical devices for scanning the horizon, cameras, notes, and a few other devices and tools.

An alternative method of studying cetaceans is through examination of dead carcasses that wash up on the shore. If properly collected and stored, these carcasses can provide important information that is difficult to obtain in field studies.

Identifying Individuals

In recent decades, methods of identifying individual cetaceans have enabled accurate population counts and insights into the lifecycles and social structures of various species.

One such successful system is photo-identification. This system was popularized by Michael Bigg, a pioneer in modern orca (killer whale) research. During the mid-1970s, Bigg and Graeme Ellis photographed local orcas in the British Columbian seas. After examining the photos, they realized they could recognize certain individual whales by looking at the shape and condition of the dorsal fin, and also the shape of the saddle patch. These are as unique as a human fingerprint; no one animal's looks is exactly like another's. After they could recognize certain individuals, they found that the animals travel in stable groups called pods. Researchers use photo identification to identify specific individuals and pods.

The photographic system has also worked well in humpback whale studies. Researchers use the color of the pectoral fins and color of the fluke to identify individuals. Scars from orca attacks found on the flukes of humpbacks are also used in identification.

REFERENCES

1. Liddell, Henry George; Scott, Robert. "δελφίς». A Greek-English Lexicon. Perseus Digital Library.
2. Dolphin. The American Heritage Dictionary of the English Language (Fourth ed.). Dictionary.com. Retrieved December 17, 2006.
3. delphinus, Charlton T. Lewis, Charles Short, A Latin Dictionary, on Perseus Digital Library
4. "mereswine - definition and meaning of mereswine at Wordnik.com". www.wordnik.com. Retrieved 2010-06-28.

5. Aquatic Life of the World. Marshall Cavendish. 1 November 2000. p. 652. ISBN 978-0-7614-7170-7. Retrieved 23 July 2013.
6. Walker, Sally M. (November 2007). Dolphins. Lerner Publications. pp. 6, 30. ISBN 978-0-8225-6767-7.
7. Andrew J. Read (1999). Porpoises. Voyageur Press. p. 13. ISBN 978-0-89658-420-4. Retrieved 23 July 2013.
8. "Coryphaena hippurus". FishBase. Retrieved 15 January 2016.
9. Stephen Leatherwood (1988). Whales, Dolphins, and Porpoises of the Eastern North Pacific and Adjacent Arctic Waters: A Guide to Their Identification. Courier Dover Publications. p. 6. ISBN 978-0-486-25651-1. Retrieved 23 July 2013.
10. Ron Hirschi (April 2002). Dolphins. Marshall Cavendish. p. 7. ISBN 978-0-7614-1443-8. Retrieved 23 July 2013.
11. Stephanie Nowacek; Douglas Nowacek (2006). Discovering Dolphins. Voyageur Press. pp. 5, 9. ISBN 978-0-7603-2561-2. Retrieved 23 July 2013.
12. Leatherwood, S.; Prematunga, W.P.; Girton, P.; McBrearty, D.; Ilangakoon, A.; McDonald, D (1991). Records of ‹blackfish' (killer, false killer, pilot, pygmy killer, and melon-headed whales) in the Indian Ocean Sanctuary, 1772-1986 in Cetaceans and cetacean research in the Indian Ocean Sanctuary. UNEP Marine Mammal Technical Report. pp. 33–65. ASIN B00KX9I8Y8.
13. "Style guide, animal names". Dictionary.com. Retrieved November 4, 2007.
14. "New Dolphin Species Discovered in Australia". September 15, 2011.
15. "Dolphin Safari sightings log". 2006. Retrieved December 17, 2006.
16. Texas Tech University (1997). "Mammals of Texas – Rough-toothed Dolphin". Retrieved December 8, 2006.
17. Robin's Island. "Dolphins at SeaWorld California". Archived from the original on 2011-07-23. Retrieved December 17, 2006.
18. Denise L. Herzing; Kelly Moewe; Barbara J. Brunnick (2003). "Interspecies interactions between Atlantic spotted dolphins, Stenella frontalis and bottlenose dolphins, Tursiops truncatus, on Great Bahama Bank, Bahamas" (PDF). Archived from the original (PDF) on June 14, 2007. Retrieved December 17, 2006.
19. Lee, Jeanette J. (April 15, 2005). "Whale-Dolphin Hybrid Has Baby ‹Wholphin'". Livescience.com. Associated Press. Archived from the original on April 17, 2005. Retrieved March 2, 2013.
20. Northeastern Ohio Universities Colleges of Medicine and Pharmacy. "Whales Descended From Tiny Deer-like Ancestors". ScienceDaily. Retrieved 21 December 2007.
21. Dawkins, Richard (2004). The Ancestor's Tale, A Pilgrimage to the Dawn of Life. Houghton Mifflin. ISBN 0-618-00583-8.
22. Thewissen, J. G. M.; Williams, E. M. (1 November 2002). "THE EARLY RADIATIONS OF CETACEA (MAMMALIA): Evolutionary Pattern and Developmental Correlations". Annual Review of Ecology and Systematics. 33

(1): 73–90. doi:10.1146/annurev.ecolsys.33.020602.095426.

23. “Introduction to Cetacea: Archaeocetes: The Oldest Whales”. University of Berkeley. Retrieved 25 July 2015.
24. Thewissen, J. G. M.; Cooper, L. N.; Clementz, M. T.; Bajpai, S.; Tiwari, B. N. (2007). “Whales originated from aquatic artiodactyls in the Eocene epoch of India” (PDF). Nature. 450 (7173): 1190–1194. Bibcode:2007Natur.450.1190T. doi:10.1038/nature06343. PMID 18097400.
25. Fahlke, Julia M.; Gingerich, Philip D.; Welsh, Robert C.; Wood, Aaron R. (2011). “Cranial asymmetry in Eocene archaeocete whales and the evolution of directional hearing in water”. PNAS. 108 (35): 14545–14548. Bibcode:2011PNAS..10814545F. doi:10.1073/pnas.1108927108. PMID 21873217.
26. “More DNA Support for a Cetacea/Hippopotamidae Clade: The Blood-Clotting Protein Gene y-Fibrinogen”. BBC News. 8 May 2002. Retrieved 20 August 2006.
27. “New Dawn”. Walking with Prehistoric Beasts. 2002. Discovery Channel.
28. Rose, Kenneth D. (2001). “The Ancestry of Whales”(PDF). 239. University of Washington: 2216–2217.
29. Bebej, R. M.; ul-Haq, M.; Zalmout, I. S.; Gingerich, P. D. (June 2012). “Morphology and Function of the Vertebral Column in Remingtonocetus domandaensis (Mammalia, cetacea) from the Middle Eocene Domanda Formation of Pakistan”. Journal of Mammalian Evolution. 19 (2): 77–104. doi:10.1007/S10914-011-9184-8.
30. Lovett, Richard A. (8 November 2006). “Dolphin With Four Fins May Prove Terrestrial Origins”. National Geographic. Retrieved 27 July 2012.
31. Gatesy, John (3 February 1997). “whales’ closest relative” (PDF). University of Arizona. Retrieved 29 August 2015.
32. “The evolution of whales”. University of Berkeley. Retrieved 29 August 2015.
33. Boisserie, Jean-Renaud; Lihoreau, Fabrice; Brunet, Michel (2005). “The position of Hippopotamidae within Cetartiodactyla”. Proceedings of the National Academy of Sciences. 102 (5): 1537–1541. Bibcode:2005PNAS..102.1537B. doi:10.1073/pnas.0409518102. PMC 547867 . PMID 15677331.
34. Katherine Ralls; Sarah Mesnick. Sexual Dimorphism(PDF). pp. 1005–1011. Retrieved 29 August 2015.
35. “Cetacean Curriculum – A teacher’s guide to introducing and using whales, dolphins, & porpoises in the classroom” (PDF). American Cetacean Society. 28 November 2004. Retrieved 20 December 2013. Sound production in cetaceans is a complex phenomenon not fully understood by scientists.
36. Scholander, Per Fredrik (1940). “Experimental investigations on the respiratory function in diving mammals and birds”. Hvalraadets Skrifter. Oslo: Norske Videnskaps-Akademi. 22.
37. Klinowska, Margaret; Cooke, Justin (1991). Dolphins, Porpoises, and Whales of the World: the IUCN Red Data Book (PDF). Retrieved 29 August 2015.
38. Stevens, C. Edward; Hume, Ian D. (1995). Comparative Physiology of the

Vertebrate Digestive System. Cambridge University Press. p. 317.

39. "Beluga Whale: The White Melon-headed Creature Of The Cold". Yellowmagpie.com. 27 June 2012. Retrieved 12 August 2013.
40. "About Whales". Whalesalive.org.au. 26 June 2009. Retrieved 12 August 2013.
41. "Boto (Amazon river dolphin) Inia geoffrensis". American Cetacean Society. 2002. Retrieved 12 September 2015.
42. Cozzi, Bruno Cozzi; Mazzario, Sandro; Podestà, Michela; Zotti, Alessandro (2009). "Diving Adaptations of the Cetacean Skeleton" (PDF). Open Zoology Journal. 2: 24–32. doi:10.2174/1874336600902010024.
43. Norena, S. R.; Williams, T. M. (2000). "Body size and skeletal muscle myoglobin of cetaceans: adaptations for maximizing dive duration". Science Direct. 126 (2): 181–191. doi:10.1016/s1095-6433(00)00182-3.
44. Nummela, Sirpa; Thewissen, J.G.M; Bajpai, Sunil; Hussain, Taseer; Kumar, Kishor (2007). "Sound transmission in archaic and modern whales: Anatomical adaptations for underwater hearing". The Anatomical Record. 290 (6): 716–733. doi:10.1002/ar.20528. PMID 17516434.
45. Thomas, Jeanette A.; Kastelein, Ronald A., eds. (2002). Sensory Abilities of Cetaceans: Laboratory and Field Evidence. 196. doi:10.1007/978-4899-0858-2. ISBN 978-1-4899-0860-5.
46. Thewissen, J. G. M. (2002). "Hearing". In Perrin, William R.; Wirsig, Bernd; Thewissen, J.G.M. Encyclopedia of Marine Mammals. Academic Press. pp. 570–2. ISBN 0-12-551340-2.
47. Ketten, Darlene R. (1992). "The Marine Mammal Ear: Specializations for Aquatic Audition and Echolocation". In Webster, Douglas B.; Fay, Richard R.; Popper, Arthur N. The Evolutionary Biology of Hearing (PDF). Springer. pp. 725–727. Retrieved March 2013. Check date values in: |access-date= (help)
48. Jeanette A. Thomas; Ronald A. Kastelein., eds. (1990). A Proposed Echolocation Receptor for the Bottlenose Dolphin (Tursiops truncatus): Modelling the Receive Directivity from Tooth and Lower Jaw Geometry NATO ASI Series A: Sensory Abilities of Cetaceans. 196. NY: Plenum. pp. 255–267. ISBN 978-0-306-43695-6.
49. Bjorn Mauck; Ulf Eysel; Guide Dehnhardt (2000). "Selective heating of vibrissal follicles in seals (Phoca vitulina) and dolphins (Sotalia fluviatilis guianensis)" (PDF). Retrieved March 11, 2007.
50. Harley, Heidi E.; DeLong, Caroline M. (2008). "Echoic Object Recognition by the Bottlenose Dolphin" (PDF). Comparative Cognition & Behavior Reviews. 3: 45–65. doi:10.3819/ccbr.2008.30003.
51. Stepanek, Laurie (May 19, 1998). "Amazon River Dolphin (Inia geoffrensis)". Texas Marine Mammal Stranding Network. Archived from the original on February 6, 2007. Retrieved November 20, 2013.
52. Mass, Alla M.; Supin, Alexander, Y. A. (21 May 2007). "Adaptive features of aquatic mammals' eyes". Anatomical Record. 290 (6): 701–715. doi:10.1002/ar.20529.

53. Davidson College, biology department (2001). "Bottlenose Dolphins – Altruism". Archived from the original on January 6, 2010. Retrieved March 12, 2008.
54. Lilley, Ray (2008-03-12). "Dolphin Saves Stuck Whales, Guides Them Back to Sea". National Geographic. Associated Press. Retrieved July 8, 2012.
55. "Dolphins save swimmers from shark". CBC News. 2004-11-24. Retrieved March 11, 2007.
56. Celizic, Mike (2007-11-08). "Dolphins save surfer from becoming shark's bait". MSNBC. Retrieved 2007-11-08.
57. Dolphin Mysteries: Unlocking the Secrets of Communication. Yale University Press. 2008. p. 83. ISBN 978-0-300-12112-4. Retrieved 16 March 2014.
58. Rowan Hooper (2005). "Dolphins teach their children to use sponges". Retrieved December 17, 2006.
59. Nic Fleming, Science correspondent for the Telegraph (December 5, 2007). "Dolphins woo females with bunches of weeds". The Daily Telegraph. London. Retrieved February 11, 2008.
60. Lilley R., 2008. Dolphin Saves Stuck Whales, Guides Them Back to Sea. http://news.nationalgeographic.com/news/2008/03/080312-AP-dolph-whal.html. The Associated Press. National Geographic News. retrieved on 24-05-2014
61. http://www.bbc.co.uk/nature/21146455
62. George Johnson. "Is Flipper A Senseless Killer?". Retrieved December 17, 2006.
63. William F. Perrin; Bernd Wursig; J. G.M. Thewissen (26 February 2009). Encyclopedia of Marine Mammals. Academic Press. ISBN 978-0-08-091993-5. Retrieved 28 June 2013.
64. Carol J. Howard (1 December 1995). Dolphin Chronicles. Random House Digital, Inc. pp. 129–. ISBN 978-0-553-37778-1. Retrieved 26 November 2012.
65. Bernd G. Würsig; Bernd Wursig, Melany Wursig (2010). The Dusky Dolphin: Master Acrobat Off Different Shores. Academic Press. pp. 156–. ISBN 978-0-12-373723-6. Retrieved 26 November 2012.
66. Edward F. Gibbons Jr.; Barbara Susan Durrant; Jack Demarest (1995). Conservation Endangered Spe: An Interdisciplinary Approach. SUNY Press. pp. 435–. ISBN 978-0-7914-1911-3. Retrieved 26 November 2012.
67. Silva Jr, J.; Silva, F.; Sazima, I. (2005). "Rest, nurture, sex, release, and play: diurnal underwater behaviour of the spinner dolphin at Fernando de Noronha Archipelago, SW Atlantic - Mating behaviour" (PDF). Archived from the original (PDF) on March 16, 2014. Retrieved 16 March 2014.
68. CMS Sotalia fluviatilis - Reproduction, article retrieved 16 March 2013.
69. Marinebio.org - Orcas (Killer Whales), Orcinus orca - Life History, article retrieved 16 March 2014.
70. Mark Simmonds, Whales and Dolphins of the World, New Holland Publishers (2007), Ch. 1, p. 32 ISBN 1-84537-820-2.
71. W. Perrin; S. Reilly (1984). Reproductive Parameters of Dolphins.

72. Volker Sommer; Paul L. Vasey (2006). "Chapter 4". Homosexual Behaviour in Animals – an Evolutionary perspective.

73. Bruce Bagemihl (1999). Biological Exuberance – Animal Homosexuality and Natural Diversity.

74. https://books.google.co.uk/books?id=tc3NGR55OsUC&lpg=PA97&dq=autocunnilingus&pg=PA98&hl=en#v=onepage&q&f=false

75. Scott, Erin; Mann, Janet; Watson-Capps, Jana; Sargeant, Brooke; Connor, Richard (2005). "Aggression in bottlenose dolphins: Evidence for sexual coercion, male-male competition, and female tolerance through analysis of tooth-rake marks and behaviour". Behaviour. 142 (1): 21–44. doi:10.1163/1568539053627712.

76. R. Connor; R. Smolker; A. Richards (1992). Two levels of alliance formation among male bottlenose dolphins.

77. http://www.huffingtonpost.com/2012/09/06/frisky-dolphin-tries-to-get-busy-divers_n_1862148.html

78. Amy Samuels; Lars Bejder; Rochelle Constantine; Sonja Heinrich (2003). "chapter 15 Marine Mammals: Fisheries, Tourism and Management Issues". Cetaceans that are typically lonely and seek human company (PDF). pp. 266–268. Retrieved December 17, 2006.

79. Dolphin's diet Archived September 27, 2013, at the Wayback Machine.

80. U.S. Department of Commerce, National Oceanic and Atmospheric Administration, National Marine Fisheries Service. "Coastal Stock(s) of Atlantic Bottlenose Dolphin: Status Review and Management Proceedings and Recommendations from a Workshop held in Beaufort, North Carolina, 13 September 1993 – 14 September 1993"(PDF). pp. 56–57.

81. Gregory K. Silber, Dagmar Fertl (1995) – Intentional beaching by bottlenose dolphins (Tursiops truncatus) in the Colorado River Delta, Mexico.

82. M.B. Santos; R. Fernández; A. López; J.A. Martínez; G.J. Pierce (2007). "Variability in the diet of bottlenose dolphin, Tursiops truncatus, in Galician waters, north-western Spain, 1990–2005". Journal of the Marine Biological Association of the UK. 87: 231. doi:10.1017/S0025315407055233.

83. "Brazil's sexiest secret". The Daily Telegraph. London. 2006-03-08. Archived from the original on 2008-03-13. Retrieved March 11, 2007.

84. Moti Nissani (2007). "Bottlenose Dolphins in Laguna Requesting a Throw Net Supporting material for Dr. Nissani's presentation at the 2007 International Ethological Conference". Retrieved February 13, 2008.

85. Viegas, Jennifer (2011). "Dolphins Talk Like Humans". Discovery News. Retrieved September 19, 2011.

86. Au, W. W. L. (1993). The Sonar of Dolphins. New York: Springer. ISBN 978-3-540-97835-0.

87. "Dolphins ‹call each other by name'". BBC News. July 22, 2013.

88. Janik, Vincent; Laela Sayigh (7 May 2013). "Communication in bottlenose dolphins: 50 years of signature whistle research". Journal of Comparative

Physiology. 199 (6): 479–489. doi:10.1007/s00359-013-0817-7.

89. "Marine Mammal vocalizations: language or behavior?". August 24, 2013.
90. "Dolphins keep lifelong social memories, longest in a non-human species". August 24, 2013.
91. Emily T. Griffiths (2009). "Whistle repertoire analysis of the short beaked Common Dolphin, Delphinus delphis, from the Celtic Deep and the Eastern and the Eastern Tropical Pacific Ocean", Master's Thesis, School of Ocean Sciences Bangor University
92. Melba C. Caldwell Et Al. – Statistical Evidence for Signture Whistles in the Spotted Dolphin, Stenella plagiodon.
93. Melba C. Caldwell Et Al. – Statistical Evidence for Signture Whistles in the Pacific Whitesided Dolphin, Lagenorhynchus obliquidens.
94. Rüdiger Riesch Et Al. – Stability and group specificity of stereotyped whistles in resident killer whales, Orcinus orca, off British Columbia.
95. Quick, Nicola; Vincent Janik (2008). "Whistle Rates of Wild Bottlenose Dolphins (Tursiops truncatus): Influences of Group Size and Behavior". Journal of Comparative Psychology. 122 (3): 305–311. doi:10.1037/0735-7036.122.3.305.
96. Weihs, D. (2002). "Dynamics of Dolphin Porpoising Revisited". Integrative and Comparative Biology. 42 (5): 1071–1078. doi:10.1093/icb/42.5.1071.
97. David Lusseau (2006), Why do dolphins jump? Interpreting the behavioural repertoire of bottlenose dolphins (Tursiops sp.) in Doubtful Sound, New Zealand
98. Corey Binns – LiveScience (2006), How Dolphins Spin, and Why, article retrieved 8 September 2013.
99. Robin D. Paulos (2010), Play in Wild and Captive Cetaceans
100. Brenda McCowan; Lori Marino; Erik Vance; Leah Walke; Diana Reiss (2000). "Bubble Ring Play of Bottlenose Dolphins (Tursiops truncatus):Implications for Cognition" (PDF). Journal of Comparative Psychology. 114 (1): 98–106. doi:10.1037/0735-7036.114.1.98.
101. Mark H. Deakos et al. (2010), Two Unusual Interactions Between a Bottlenose Dolphin (Tursiops truncatus) and a Humpback Whale (Megaptera novaeangliae) in Hawaiian Waters.
102. Cathy Hayes for Irish Central (2011), Amazing footage of a dog playing with a dolphin off the coast of Ireland, article retrieved 8 September 2013.
103. Siebert, Charles (8 July 2009). "Watching Whales Watching Us". New York Times Magazine. Retrieved 29 August 2015.
104. Watson, K.K.; Jones, T. K.; Allman, J. M. (2006). "Dendritic architecture of the Von Economo neurons". Neuroscience. 141 (3): 1107–1112. doi:10.1016/j.neuroscience.2006.04.084. PMID 16797136.
105. Allman, John M.; Watson, Karli K.; Tetreault, Nicole A.; Hakeem, Atiya Y. (2005). "Intuition and autism: a possible role for Von Economo neurons". Trends Cogn Sci. 9 (8): 367–373. doi:10.1016/j.tics.2005.06.008. PMID 16002323.

106. Hof, Patrick R.; Van Der Gucht, Estel (2007). "Structure of the cerebral cortex of the humpback whale, Megaptera novaeangliae (Cetacea, Mysticeti, Balaenopteridae)". The Anatomical Record. 290 (1): 1–31. doi:10.1002/ar.20407. PMID 17441195.

107. Moore, Jim. "Allometry". University of California San Diego. Retrieved 9 August 2015.

108. "How smart are they?". Orlando Sentinel. March 7, 2010.

109. Fields, R. Douglas. "Are whales smarter than we are?". Scientific American. Retrieved 9 August 2015.

110. "Elephant Self-Awareness Mirrors Humans". Live Science. 30 October 2006. Retrieved 29 August 2015.

111. Derr, Mark. "Mirror test". New York Times. Retrieved 3 August 2015.

112. Marten, Ken; Psarakos, Suchi (June 1995). "Using Self-View Television to Distinguish between Self-Examination and Social Behavior in the Bottlenose Dolphin (Tursiops truncatus)". Consciousness and Cognition. 4 (2): 205–224. doi:10.1006/ccog.1995.1026.

113. Mukhametov, L. M.; Supin, A. Ya. (1978). "Sleep and vigil in dolphins". Marine mammals. Moscow: Nauka.

114. Mukhametov, Lev (1984). "Sleep in marine mammals". Experimental Brain Research. Experimental Brain Research Supplementum. 8 (suppl.): 227–238. doi:10.1007/978-3-642-69554-4_17. ISBN 978-3-642-69556-8.

115. Dallas Grasby (1994). "Excerpts from "Sleep in marine mammals", L.M. Mukhametov". Experimental Brain Research. 8 (suppl.). Retrieved February 11, 2008.

116. McCormick, James G. (2007). "Behavioral Observations of Sleep and Anesthesia in the Dolphin: Implications for Bispectral Index Monitoring of Unihemispheric Effects in Dolphins". Anesthesia & Analgesia. 104 (1): 239–241. doi:10.1213/01.ane.0000250369.33700.eb.

117. BBC (2008-02-20). "Sperm whales caught ‹cat napping'". BBC News. Retrieved April 9, 2008.

118. Martin, G. Neil; Carlson, Neil R.; Buskist, William (1997). Psychology, third edition. Pearson Allyn & Bacon. p. 383. ISBN 978-0-273-71086-8.

119. Michael R. Heithaus (1999), Predator-prey and competitive interactions between sharks (order Selachii) and dolphins (parvorder Odontoceti): a review

120. Emine Sinmaz for the Daily Mail (2013), Eight-ton orca leaps 15ft into the air to finally capture dolphin he wanted for dinner after two-hour chase, article retrieved 8 September 2013.

121. Nadia Drake for WIRED (2013), Photographer Captures Stunning Killer Whale Attack on Dolphin, article retrieved 8 September 2013.

122. Bossart, G. D. (2007). "Emerging diseases in marine mammals: from dolphins to manatees" (PDF). Microbe. 2 (11): 544–549.

123. Woodard, J. C.; Zam, S. G.; Caldwell, D. K.; Caldwell, M. C. (1969). "Some Parasitic Diseases of Dolphins" (PDF). Veterinary Pathology. 6 (3): 257–272. doi:10.1177/030098586900600307.

124. Bellière, E. N.; Esperón, F.; Sánchez-Vizcaíno, J. M. (2011). "Genetic comparison among dolphin morbillivirus in the 1990–1992 and 2006–2008 Mediterranean outbreaks". Infection, Genetics and Evolution. 11 (8): 1913–1920. doi:10.1016/j.meegid.2011.08.018. PMID 21888991.

125. Jane J. Lee for National Geographic (2013), What's Killing Bottlenose Dolphins? Experts Discover Cause., article retrieved 4 December 2013.

126. Guardo, G. D.; Marruchella, G.; Agrimi, U.; Kennedy, S. (2005). "Morbillivirus Infections in Aquatic Mammals: A Brief Overview". Journal of Veterinary Medicine Series A. 52 (2): 88–93. doi:10.1111/j.1439-0442.2005.00693.x. PMID 15737178.

127. Stone, B. M.; Blyde, D. J.; Saliki, J. T.; Blas-Machado, U.; Bingham, J.; Hyatt, A.; Wang, J.; Payne, J.; Crameri, S. (2011). "Fatal cetacean morbillivirus infection in an Australian offshore bottlenose dolphin (Tursiops truncatus)". Australian Veterinary Journal. 89 (11): 452–457. doi:10.1111/j.1751-0813.2011.00849.x. PMID 22008125.

128. US National Marine Mammal Foundation (2010-02-18). "Scientists Find Clues in Dolphins to Treat Diabetes in Humans". Archived from the original on 2010-04-11. Retrieved February 20, 2010.

129. Georgetown University Medical Center. "Dolphins' "Remarkable" Recovery from Injury Offers Important Insights for Human Healing". Retrieved July 26, 2011.

130. Viegas, Jen (2014-08-14). "Shark vs. Dolphin Battles Can Have Surprising Outcomes". Seeker - Science. World. Exploration. Seek for yourself. Retrieved 2016-11-27.

131. Douglas Williams for Shanghai Daily (December 4, 2006). "Yangtze dolphin may be extinct". Retrieved December 9, 2006.[dead link]

132. "TED: Stephen Palumbi: Following the mercury trail". Youtube.com. June 30, 2010. Retrieved 2011-07-25.

133. Clover, Charles. (2004). The End of the Line: How overfishing is changing the world and what we eat. Ebury Press, London. ISBN 978-0-09-189780-2.

134. Díaz López, Bruno; Bernal Shirai, J.A. (2006). "Bottlenose dolphin (Tursiops truncatus) presence and incidental capture in a marine fish farm on the north-eastern coast of Sardinia (Italy)". Journal of Marine Biological Association UK. 87: 113–117. doi:10.1017/S0025315407054215.

135. Díaz López, Bruno (2006). "Interactions between Mediterranean bottlenose dolphins (Tursiops truncatus) and gillnets off Sardinia, Italy". ICES Journal of Marine Science. 63 (5): 946–951. doi:10.1016/j.icesjms.2005.06.012.

136. Matsutani, Minoru (2009-09-23). "Details on how Japan's dolphin catches work". Japan Times. p. 3.

137. Johnston, Eric (2009-09-23). "Mercury danger in dolphin meat". Japan Times.

p. 3.

138. "Time Of Truth For US Dolphin Safe Logos". atuna. Retrieved 29 August 2015.
139. CBC news (2003-10-09). "Navy sonar may be killing whales, dolphins". CBC News. Retrieved October 27, 2008.
140. "Npower renewables Underwater noise & vibration, section 9.4" (PDF). Archived from the original (PDF) on 2007-07-22. Retrieved October 27, 2008.
141. Hentea, Călin (2007). Brief Romanian Military History. Maryland, USA: Scarecrow Press. p. 91. ISBN 978-0-8108-5820-6.
142. Ali, Arif (1996). Barbados: Just Beyond Your Imagination. Hertford, UK: Hansib Publications. p. 16. ISBN 978-1-870518-54-3.
143. "Taras". Archived from the original on 2007-02-08. Retrieved 3 October 2010.
144. Eyers, Jonathan (2011). Don't Shoot the Albatross!: Nautical Myths and Superstitions. A&C Black, London, UK. ISBN 978-1-4081-3131-2.
145. Herodotus I.23; Thucydides I.128, 133; Pausanias iii.25, 4
146. Vijay Singh (1994). The River Goddess. London. ISBN 978-1-85103-195-5.
147. Atlantis: The Palm, Dubai. "Swim With the Dolphins in Dubai". atlantisthepalm.com. Retrieved 11 March 2016.
148. Dubai Dolphinarium. "The Dubai Dolphinarium". dubaidolphinarium.ae. Retrieved 11 March 2016.
149. Rose, Naoimi; Parsons, E.C.M.; Farinato, Richard. The Case Against Marine Mammals in Captivity (PDF) (4 ed.). Humane Society of the United States. pp. 13, 42, 43, 59.
150. Sean B. Carroll. "Remarkable creatures". Retrieved 6 January 2016.
151. "Orcas in Captivity - A look at killer whales in aquariums and parks]". 23 November 2009. Retrieved 6 September 2015.
152. NMFS 2005, pp. 43–44.
153. india.gov. "National Aquatic Animal". india.gov.
154. Rose, N. A. (2011). "Killer Controversy: Why Orcas Should No Longer Be Kept in Captivity" (PDF). Humane Society International and the Humane Society of the United States. Retrieved December 21, 2014.
155. "Whale Attack Renews Captive Animal Debate". CBS News. March 1, 2010. Retrieved 6 September 2015.
156. Susan Jean Armstrong. Animal Ethics Reader. ISBN 978-0-415-27589-7.
157. William Broad (1999-07-06). "An article on the aggressive nature of dolphins". Fishingnj.org. Retrieved 2011-07-25.
158. Zimmerman, Tim (2011). "The Killer in the Pool". The Best American Sampler 2011. Houghton Mifflin Harcourt. p. 336.
159. "Corpse Is Found on Whale". New York Times. July 7, 1999. Retrieved September 11, 2011.
160. "SeaWorld trainer killed by killer whale". CNN. February 25, 2010. Retrieved

March 11, 2010.

161. Pilkington, Ed (2010-02-25). “Killer whale Tilikum to be spared after drowning trainer by ponytail”. The Guardian. London.
162. “Killer whale bumps but doesn’t bite boy”. Juneau Empire – Alaska’s Capital City Online Newspaper. 2005-08-19. Archived from the original on 2010-04-11. Retrieved 2011-07-25.
163. Wong, David. “The 6 Cutest Animals That Can Still Destroy You”. Cracked.com. Retrieved 2011-07-25.
164. “Male Dolphin Kills Man”. Science-frontiers.com. Sep–Oct 1995. Retrieved 2011-07-25.
165. “LONE DOLPHINS – FRIEND OR FOE?”. BBC. 2002-09-09. Retrieved 2012-02-05.
166. Thomas I. White (2007). In defense of dolphins: the new moral frontier. Blackwell Public Philosophy Series. ISBN 978-1-4051-5778-0.
167. Jonathan Leake and Helen Brooks for The Sunday Times (2010), Scientists say dolphins should be treated as ‹non-human persons’, article retrieved January 4, 2010.
168. Land, Graham (29 July 2013). Dolphin rights: The world should follow India’s lead. Asiancorrespondent.com. Hybrid News Ltd. Accessed on 29 July 2013.
169. Policy on establishment of dolphinarium
170. PBS – Frontline. “The Story of Navy dolphins”. Retrieved June 8, 2008.
171. “Iran buys kamikaze dolphins”. BBC News. 2000-03-08. Retrieved June 7, 2008.
172. Antonioli, C; Reveley, MA (2005). “Randomised controlled trial of animal facilitated therapy with dolphins in the treatment of depression”. BMJ (Clinical research ed.). 331 (7527): 1231. doi:10.1136/bmj.331.7527.1231. JSTOR 25455488. PMC 1289317 . PMID 16308382.
173. Biju Basil; Maju Mathews (2005). “Methodological concerns about animal-facilitated therapy with dolphins”. BMJ. 331 (7529): 1407. doi:10.1136/bmj.331.7529.1407. PMC 1309662 . PMID 16339258.
174. Lori Marino; Scott O. Lilienfeld (2007). “Dolphin-Assisted Therapy: more flawed data and more flawed conclusions”(PDF). Anthrozoös. 20 (3): 239–49. doi:10.2752/089279307X224782. Archived from the original (PDF) on 2008-02-28. Retrieved 2008-02-20.
175. Matsutani, Minoru (September 23, 2009). “Details on how Japan’s dolphin catches work”. Japan Times. p. 3.
176. McCurry, Justin (2009-09-14). “Dolphin slaughter turns sea red as Japan hunting season returns”. The Guardian. London. Retrieved 24 May 2010.
177. Hall, Kevin G. (2003). “Dolphin meat widely available in Peruvian stores: Despite protected status, ‹sea pork’ is popular fare”. The Seattle Times.
178. [Dolphin in Miso Vegetable Stew]. Cookpad (in Japanese). 2010. Retrieved 16 May 2011.

179. World Health Organisation / United Nations Environment Programme DTIE Chemicals Branch (2008). "Guidance for identifying populations at risk from mercury exposure" (PDF). p. 36. Retrieved 29 August 2013.

CHAPTER 7

Paleobiology

Chapter Contents

- **Introduction**
- **Sub-Disciplines**
- **Japanese Primatology**
- **Women in Primatology**

Paleobiology is a branch of the natural sciences which focuses on examining the origins and history of life on Earth through use of the fossil record. In this sense, it could be considered a branch of paleontology, but it also combines aspects of other scientific fields. As a specific field of study, paleobiology emerged in the 20th century, and scientific advancement has made the scope of this scientific field much wider. Many major universities offer paleobiology programs, with paleobiologists working at universities, museums, and research facilities around the world.

INTRODUCTION

Palaeozoology, also spelled as Paleozoology is the branch of paleontology, paleobiology, or zoology dealing with the recovery and identification of multicellular animal remains from geological (or even archeological) contexts, and the use of these fossils in the reconstruction of prehistoric environments and ancient ecosystems.

Definitive, macroscopic remains of these metazoans are found in the fossil record from the Ediacaran period of the Neoproterozoic era onwards, although they do not become common until the Late Devonian period in the latter half of the Paleozoic era.

Perhaps the best known macrofossils group is the dinosaurs. Other popularly known animal-derived macrofossils include trilobites, crustaceans, echinoderms, brachiopods, mollusks, bony fishes, sharks, Vertebrate teeth, and shells of numerous invertebrate groups. This is because hard organic parts, such as bones, teeth, and shells resist decay, and are the most commonly preserved and found animal fossils. Exclusively soft-bodied animals—

such as jellyfish, flatworms, nematodes, and insects—are consequently rarely fossilized, as these groups do not produce hard organic parts.

Priscacara liops

VERTEBRATE PALEOZOOLOGY

Vertebrate paleozoology refers to the use of morphological, temporal, and stratigraphic data to map vertebrate history in evolutionary theory. Vertebrates are classified as a subphylum of Chordata, a phylum used to classify species adhering to a rod-shaped, flexible body type called a notochord. They differ from other phyla in that other phyla may have cartilage or cartilage-like tissues forming a sort of skeleton, but only vertebrates possess what we define as bone.

Classes of vertebrates listed in chronological order from oldest to most recent include heterostracans, osteostracans, coelolepid agnathans, acanthodians, osteichthyan fishes, chondrichthyan fishes, amphibians, reptiles, mammals, and birds. All vertebrates are studied under standard evolutionary generalizations of behavior and life process, although there is controversy over whether population can be accurately estimated from limited fossil resources.

Evolutionary origins of vertebrates as well as the phylum Chordata have not been scientifically determined. Many believe vertebrates diverged from a common ancestor of chordates and echinoderms. This belief is well supported by the prehistoric marine creature Amphioxus. Amphioxus does not possess bone, making it an invertebrate, but it has common features with vertebrates including a segmented body and a notochord. This could imply that Amphioxus is a transitional form between an early chordate, echinoderm or common ancestor, and vertebrates.

QUANTITATIVE PALEOZOOLOGY

Quantitative paleozoology is a process of taking a census of fossil types rather than inventory. They differ in that inventory refers to a detailed log of individual fossils, whereas census attempts to group individual fossils to tally the total number of a species. This information can be used to determine which species were most dominant and which had the largest population at a time period or in a geological region.

In the early 1930s, paleontologists Chester Stock and Hildegarde Howard devised special units for quantitative paleozoology and quantitative paleontology. The first unit used, Number of Identified Species (NISP), specified exact quantity of fossils from a specific species recorded. Stock and Howard determined this unit to be problematic for quantitative purposes as an excess of a small fossil—such as teeth—could exaggerate quantity of the species. There was also an amount of confusion as to whether bone fragments should be assembled and counted as one bone or tallied individually. Stock and Howard then devised the Minimum Number of Individuals (MNI), which estimated the minimum number of animals needed to produce the fossils recorded. For example, if five scapulae from a species were found, it might be difficult to determine whether some of them were paired right and left on one individual or whether each came from a different individual, which could alter census, but it could be said that there must be at least three individuals to produce five scapulae. Five would thus be the MNI. In rare cases where enough of a collection of fossils can be assembled into individuals as to provide an accurate number of individuals, the unit used is Actual Number of Individuals, or ANI.

Another unit commonly used in quantitative paleozoology is biomass. Biomass is defined as the amount of tissue in an area or from a species. It is calculated by estimating an average weight based on similar modern species and multiplying it by the MNI. This yields an estimate of how much the entire population of a species may have weighed. Problems with this measurement include the difference in weight between youngsters and adults, seasonal weight changes due to diet and hibernation, and the difficulty of accurately estimating the weight of a creature with only a skeletal reference. It is also difficult to determine exact age of fossilized matter within a year or a decade, so a biomass might be grossly exaggerated or under exaggerated if the estimated time frame in which the fossils were alive is incorrect.

A similar measurement to biomass is meat weight. To determine meat weight, MNI is multiplied by the amount of meat an individual is thought to have provided, then multiplied by the percentage of that meat thought to be edible. This gives an estimate of "pounds of usable meat" per individual which might have been harvested by prehistoric hunters. For example, a male Wapiti has an average weight of 400 kg, and in a particular study, the MNI of Wapiti was found to be 10. This would create a biomass of 4,000 kg. If the amount of edible meat is estimated at 50 percent, this would result in a meat weight of 2,000 kg. The biggest problem with this method is the debate over "percent of usable meat." Different views on which parts of a species are edible and which are not as well as whether or not primitive butchers would have been able to access and prepare different parts have led to controversy.

CONSERVATION BIOLOGY

Paleozoological data is used in research concerning conservation biology. Conservation biology refers to biological study used for conservation, control, and preservation of various species and ecosystems. In this context, the paleozoological data used is obtained from recently deceased decomposing matter rather than prehistoric matter.

R. Lee Lyman, Professor and Chair Department of Anthropology at the University of Missouri, writes that paleozoological research can provide data such as extinction rates and causes and "benchmark" peaks and drops in population which can be used to predict future patterns and to design maximally effective methods of controlling these patterns. In addition, paleozoological data can be used to compare current to former population and distribution of a species.

REFERENCES

1. Lyman, R. Lee (2008). Quantitative Paleozoology. Cambridge: Cambridge University Press.
2. Olson, Everett C. (1971). Vertebrate Paleozoology. New York: Wiley-Interscience.
3. Lyman, R. Lee (February 2006). "Paleozoology in the Service of Conservation Biology". Evolutionary Anthropology. 15 (1): 11–19. doi:10.1002/evan.20083. Retrieved 28 March 2012.
4. "R. Lee Lyman: Zooarchaeology and Vertebrate Taphonomy". Curators of the University of Missouri. Retrieved 12 April 2012.

Index

A

African Arachnological Society (AFRAS) 60
Allometric analysis 123
American Arachnological Society (AAS) 60
American Society 192
Animal species 187
Arachnological Society of Japan (ASJ) 60
Asian Society of Arachnology (ASA) 60

B

Behavioral ecology 149
Beta taxonomy 25
Biological classification 31
Biological taxonomy 30
British Arachnological Society (BAS) 60

C

Chronological order 260

E

European Society of Arachnology (ESA) 61
evolutionarily stable strategies (ESS) 159

F

False killer whales 123, 142
Familial conflict 159
Folia Primatologica 192

H

Humanitarian values 191

I

International Code of Zoological Nomenclature (ICZN Code) 31
International Primatological Society 192
International Society of Arachnology (ISA) 60
International Union for Conservation of Nature (IUCN) 103, 108
Iranian Arachnological Society (IAS) 60

J

Journal of Human Evolution 192, 193

M

Minimum Number of Individuals (MNI) 261

N

Natural environments 183
Number of Identified Species (NISP) 261

P

Paleobiology 259
Paleozoological data 262
Partial classifications 30
Phenetic analyses 38

Primatology 181, 182, 183, 187, 191, 192, 193

Q

Quantitative paleozoology 261

S

Scientific studies 190
Self-awareness 123, 142
Social interactions 124

T

Taxonomy 23, 24, 26, 39, 40, 41
The Associate Certified Entomologist (ACE) 50

V

Vertebrate paleozoology 260

W

Western primatology stems 182